The Institute of Biology's
Studies in Biology no. 33

Animal
Flight

by Colin Pennycuick Ph.D.

Edward Arnold

© Colin Pennycuick 1972

First published 1972
by Edward Arnold (Publishers) Limited,
25 Hill Street,
London, W1X 8LL

Boards edition ISBN: 0 7131 2355 9
Paper edition ISBN: 0 7131 2356 7

Printed in Great Britain by
William Clowes & Sons, Limited
London, Beccles and Colchester

General Preface to the Series

It is no longer possible for one textbook to cover the whole field of Biology and to remain sufficiently up to date. At the same time students at school, and indeed those in their first year at universities, must be contemporary in their biological outlook and know where the most important developments are taking place.

The Biological Education Committee, set up jointly by the Royal Society and the Institute of Biology, is sponsoring, therefore, the production of a series of booklets dealing with limited biological topics in which recent progress has been most rapid and important.

A feature of the series is that the booklets indicate as clearly as possible the methods that have been employed in elucidating the problems with which they deal. Wherever appropriate there are suggestions for practical work for the student. To ensure that each booklet is kept up to date, comments and questions about the contents may be sent to the author or the Institute.

1972   INSTITUTE OF BIOLOGY
                                                    41 Queen's Gate
                                                    London, S.W.7

Preface

My aim in this book has been to show in a general way how the anatomy, physiology and performance of flying animals are related to the mechanical principles of flight. In a book of this length a certain amount of oversimplification cannot be avoided. The validity of my attempt to explain the physical principles which underlie animal flight depends, at least in part, on the level of realism of the approximations used in the argument. No doubt one or two of these are not so realistic as one might wish, because of the lack of experimental data, and I have resorted to a little judicious guesswork to fill some gaps.

The general approach which I have used has taken shape in discussions with innumerable colleagues, and has been tried out on several generations of long-suffering students. I am most grateful to them all, and I am also no less indebted to my flying friends, especially those who taught me to fly, and to soar. The errors remain, of course, exclusively my own.

Seronera, 1972   C.J.P.

Contents

Kinds of Flight 1

1.1 Flying animals

The power of sustained horizontal or climbing flight is found in members of only two phyla, the chordates and the arthropods. Among these, three living groups, the insects, birds and bats have made flight their characteristic method of locomotion, and the members of one extinct group, the pterosaurs, were certainly proficient fliers in their day. More limited powers of parachuting and gliding flight occur in all classes of vertebrates, and at least one mollusc, and are more conveniently classified in terms of the mechanical principles involved, rather than the systematic affinities of their owners.

The spectrum of animal flight can be broadly divided into three levels—(1) parachuting, (2) gliding and (3) powered flight. These categories shade into one another, and should not be thought of as sharply distinct.

1.2 Parachuting

A pure parachute is a structure which develops an aerodynamic force parallel to the direction of the air flowing past it. Such a force is called *drag* force.

Any flying object reaches equilibrium when the net aerodynamic force acting on it balances the weight which, by definition, always acts vertically downwards. A pure drag force, which acts parallel to the direction of relative air flow, can therefore only balance the weight if the object is travelling vertically downwards. It may be noted that an object descending vertically through still air at some speed V_z is aerodynamically indistinguishable from the same object stationary, with the surrounding air moving upwards past it at V_z. It is the speed of the air *relative to the object* which determines the magnitude and direction of the aerodynamic force: in the case of a vertically descending object the air appears to be moving upwards relative to it, and thus produces an upward drag force.

The magnitude of the drag force depends on three things, the effective surface area of the parachute, the speed of the relative air flow, and the density of the air. For a given weight, the bigger the parachute, the slower it falls. If the weight is W and the surface area S, the speed of fall of different parachutes of the same shape is approximately proportional to $\sqrt{W/S}$.

A parachute cannot hover motionless in still air because there is then no relative air flow and hence no upward force to balance the weight. The faster the relative air flow, the bigger the force—when a parachute is

dropped, it accelerates downwards until it is going fast enough to generate a force equal to the weight, after which it is in equilibrium and
continues downwards at a constant speed.

1.3   Gliding

Very few animals act as pure parachutes, although some symmetrical
plant propagules such as those of the dandelion may do so. Flying animals are mostly not radially symmetrical, and deflect the air flowing past
their bodies, so that the reaction from it is not parallel to the direction of
flow. This is the basis of the distinction between parachuting and
gliding.

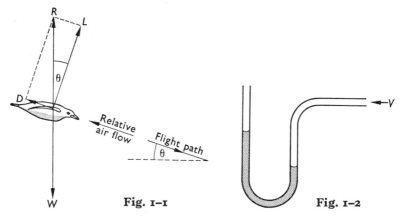

Fig. 1-1   Fig. 1-2

Fig. 1-1   (Left) A bird gliding at a steady speed descends at an angle θ to
the horizontal, and therefore 'sees' the relative air flow coming upwards
towards it at this angle. The net aerodynamic force R must balance the
weight, and therefore acts vertically upwards. It is inclined at an angle
$(\frac{1}{2}\pi - \theta)$ to the relative wind, and this is also the angle between the flight path
and the vertical. The net aerodynamic force is traditionally analysed into two
components, the drag D parallel with the relative air flow, and the lift L, at
right angles to it. The lift does not act vertically upwards in this situation.
Fig. 1-2   (Right) See text.

In a straight glide at constant speed the weight is balanced by the net
aerodynamic force, as in a steadily descending parachute. However, as
this force is now not parallel to the direction of the relative air flow, the
flight path is no longer directed vertically downwards, but slopes at an
angle to the vertical—at the same angle, in fact, as the angle between the
net aerodynamic force and the direction of the relative air flow (Fig. 1-1).
In this situation one can define the *drag* in the same way as before—as
that component of the net aerodynamic force which acts parallel to the
direction of the relative air flow, i.e., along the flight path. It is customary

to resolve the net aerodynamic force into the drag so defined, and a second component, the *lift* which acts at right angles to the direction of the relative air flow. It should be noted that the lift so defined does not necessarily act upwards—in fact, on a flapping wing, the lift on different parts can act in different directions simultaneously, because of variations in the direction of the relative air flow, caused by the wing's own motion.

It can be seen from Fig. 1–1 that the *gliding angle*, or angle between the flight path and the horizontal, is the complement of the angle between the directions of the net aerodynamic force and the relative airflow. The latter angle is usually referred to in terms of its tangent, the *lift:drag ratio* L/D, also called the 'glide ratio' or just 'L over D'. A bird in which the net aerodynamic force acted at right angles to the relative air flow (that is, $L/D=\infty$) could glide horizontally without losing height or speed, but this can never be achieved in reality. In practice there is always some drag. Flying objects with very poor glide ratios, say less than 1, are generally classified, for convenience, with parachutes.

Unlike a parachute, a gliding bird can travel at a range of different speeds, but there is a definite minimum speed, slower than which it cannot glide. As in the parachute, this minimum speed is reduced by increasing the wing area, and increased by adding weight. If weight is added near the existing centre of gravity of a particular glider, the gliding angle is not much affected, but the heavier it is, the faster it goes. Many people find this fact surprising, but the reader who owns a toy glider and a supply of Plasticene can readily verify it.

1.4   Lift and drag coefficients

The actual lift force acting on a wing may vary from a dyne or two to many tonnes, depending on whether the wing belongs to a gnat or an airliner, and when drawing comparisons between wings of different sizes flying under different conditions, it is usual to consider not the lift itself, but the *lift coefficient*. This is the ratio between the lift and a reference force, which takes into account the wing area, the air density and the speed, and is made up in the following way.

If a stream of air of density ρ blows at a speed V towards an open-ended tube connected to a manometer (Fig. 1–2), then since the air in the tube is stationary, the approaching air molecules have to be brought to a stop before they reach it. This results in a rise of pressure p in the tube, which is registered by the manometer, and is given by

$$p = \tfrac{1}{2}\rho V^2 \tag{1.1}$$

p is called the *dynamic pressure* and is characteristic of a particular air density and wind speed. When multiplied by the wing area S it gives a reference force F_r, which is characteristic of this density and speed, and also of the wing area:

$$F_r = \tfrac{1}{2}\rho V^2 S \qquad (1.2)$$

The lift coefficient C_L is defined as the ratio

$$C_L = \frac{L}{\tfrac{1}{2}\rho V^2 S} \qquad (1.3)$$

The lift coefficient, being the ratio of two forces, is dimensionless, and its value is independent of the system of units in which the forces themselves are measured.

Lift results from the downward deflection of the air passing over the wing, and the lift coefficient is closely related to the *downwash angle*, i.e., the angle through which air is deflected. For any particular shape of wing this downwash angle depends on the *angle of attack*, which is the angle between the plane of the wing and the direction of the incident air flow. For most wings the downwash angle, and also the lift coefficient, is proportional to the angle of attack over a range of $10-20°$ for the latter, but if the angle of attack is made very large the change of direction required of the incident air becomes too sharp, and the air stream

Fig. 1–3 Behaviour of aerofoil sections. The thin lines represent the paths of air molecules, approaching each section from the left. **(a)** Unstalled wing at a moderate angle of attack. **(b)** Stalled wing, showing separated flow and reduced downwash angle. **(c)** Use of a projection on the upper surface to delay the stall in a small-scale wing, by inducing the formation of a turbulent boundary layer.

breaks away from the wing surface, with a marked reduction of downwash angle and of lift coefficient. A wing in this condition is said to be *stalled* (Fig. 1–3). The maximum lift coefficient C_{Lmax} is obtained just before the stall and determines the minimum gliding speed, or *stalling speed* V_s. In a steady glide the lift approximately equals the weight, and so from equation (1.3) it follows that

$$V_s = \sqrt{\frac{2W}{\rho S C_{Lmax}}} \qquad (1.4)$$

The ratio W/S is known as the *wing loading*, and, other things being equal, the stalling speed should vary with the square root of the wing loading. Similar laws apply to other characteristic speeds, such as those for minimum power and maximum range, which will be discussed in section 2.5.

Measured maximum lift coefficients are 1.5–1.6 in gliding birds, 1.5 in a gliding bat, and 1.3 in an isolated locust forewing. Indirect evidence suggests that bird wings may be capable of lift coefficients as high as 2.8 in flapping flight. Aircraft wings with elaborate slots and flaps have achieved lift coefficients up to about 4, but it is generally easier to obtain high lift coefficients with large wings than with small ones. Separation of the flow from the upper surface, as in Fig. 1–3b, can be delayed in small wings by stimulating the formation of turbulence just above the wing surface (Fig. 1–3c). Aeromodellers achieve this by gluing a thread along the upper wing surface, and the same effect is produced in bats by the projection of the wing bones above the upper surface, and in insects by the wing 'veins'. A turbulent 'boundary layer' induced in this way should not be confused with the much larger-scale turbulence above a stalled wing, as in Fig. 1–3b—in this case the boundary layer might be laminar, but is no longer attached to the wing surface. In aircraft-sized wings the transition to turbulent flow takes place automatically at the minimum pressure point, and so no special precautions to induce turbulence are necessary.

1.5   Parachuting and gliding animals

Many groups of arboreal vertebrates have independently evolved systems for parachuting or gliding, which have obvious uses in travel from tree to tree, and in escaping from predators (Fig. 1–4). At the crudest level the 'flying frogs' of the genus *Rhacophorus* parachute by spreading out their large webbed feet. The gliding lizard *Draco volans* has quite an efficient gliding wing, based on a fold of skin supported on elongated ribs, which can be folded against the sides of the body when not in use. This system has the advantage of being independent of the legs, all four of which are left free for normal locomotion.

Among the mammals a fold of skin (or patagium) stretched between the fore and hind legs has been evolved independently as a gliding wing in at least four arboreal groups. The flying squirrels of the family Sciuridae and the scaly-tails (Anomaluridae) are both rodents, and superficially similar in structure and habits, as are the marsupial flying phalangers of the family Phalangeridae (see Plate 1). The two species of cobegos (misnamed 'flying lemurs'), *Cynocephalus*, constitute the sole genus of the order Dermoptera, and differ from other gliding mammals in that the patagium includes the fingers and toes (rather than ending at the wrists and ankles) and also extends to the chin and tail.

In the flying fishes of the family Exocoetidae the pectoral fins have been expanded into wings, which enable them to glide for a considerable distance after leaping from the surface of the water, thus baffling their submarine pursuers. There is also a flying squid which uses similar tactics.

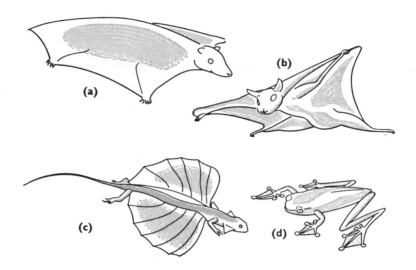

Fig. 1–4 Parachuting and gliding animals. (**a**) The cobego or 'flying lemur' *Cynocephalus volans* (Dermoptera) of the Philippines. (**b**) A North American flying squirrel *Glaucomys volans* (Rodentia: Sciuridae). (**c**) The East Indian gliding lizard *Draco volans* (Lacertilia: Agamidae). (**d**) A parachuting frog from Borneo, *Rhacophorus dulitensis* (Anura: Polypedatidae).

1.6   Powered flight

Because there is always some drag, gliding flight cannot be horizontal, except briefly, at the expense of losing speed. To fly horizontally at a steady speed, a *thrust* force T must be provided, equal and opposite to the drag, so producing the balanced situation shown in Fig. 1–5. In this special case of level flight, $T = D$, and the *power* required P is given by

$$P = TV = DV \qquad (1.5)$$

This power is supplied by the flight muscles, and can be accounted for in terms of the *rates* of consumption of fuel and oxygen.

The thrust need not, of course, necessarily equal the drag. If $T > D$, the flight path is inclined upwards, and the bird climbs, while if $T < D$, it descends. Descending flight shades into another special case where

$T=0$ ('engine off'), which is the gliding case considered in the last section.

The faculty of steady horizontal flight is an evolutionary achievement of special importance, and is sometimes referred to as 'true' flight (all

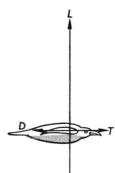
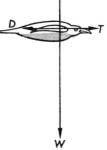

Fig. 1-5 Simplified equilibrium diagram for horizontal powered flight.

other kinds being presumably considered 'false'). This is misleading, however, and as we shall see the most advanced fliers, the soaring birds, spend most of their airborne time gliding, some of them being only marginally capable of flying horizontally.

Mechanics of Powered Flight 2

2.1 Analysis of power requirements

A flying bird is very complicated aerodynamically, and it is not profitable to try to describe everything that goes on in the air around it in full detail. However, an understanding of the main processes involved in flight is very helpful in assessing the capabilities of a particular kind of flying animal, and also in understanding the relation between the animal's anatomy and the sort of flight to which it is adapted. As far as the animal is concerned, in order to fly it is obliged to do mechanical work with its muscles. The *rate* at which it has to do work is the mechanical power required to fly, and this is reflected in the rates at which fuel and oxygen are consumed. The purpose of this chapter is to show in a general way how the power required in horizontal flight is related to the animal's weight and wing span, and to its forward speed. A more detailed justification of the results given here is given by PENNYCUICK (1969).

Although the power all comes from the same muscles, it is required for three distinct purposes, and the amounts of power needed for each are affected differently by evolutionary changes in anatomy. Also, in a particular animal, they vary differently with changes in the forward speed. The three main purposes for which the power is required are:

(1) To support the weight. This component is called the *induced power*.
(2) To overcome the profile drag of the body (*parasite power*).
(3) To overcome the profile drag of the wings (*profile power*).

2.2 Induced power

A stationary object is said to be 'at rest', and a land animal lying motionless, or a fish with neutral buoyancy, can indeed remain where it is without having to do any muscular work. In contrast, a bird has to work very hard indeed to hover 'at rest' in mid-air, and hovering is considerably more strenuous than flying forwards at some moderate speed.

The force which holds a bird up is a reaction produced by continuously accelerating air downwards, and must equal the weight at any speed. Beginning with the case of zero speed (hovering), the bird beats its wings to and fro—more or less horizontally—sweeping out an area known as the *wing disc* (Fig. 2–1). A continuous stream of air is sucked from above and accelerated downwards through the disc (Fig. 2–1b), passing through it at some speed called the *induced velocity*. The magnitude of the upward *force* developed is equal to the rate at which downward momentum is imparted to the air, and is proportional both to the induced velocity and to the *mass flow* through the wing disc—that is, the rate, in mass per unit time, at which air passes through the disc. The *power* required, on the other hand, is proportional only to the induced velocity, and does not depend on the mass flow.

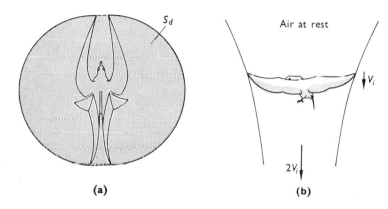

(a)   (b)

Fig. 2–1 In a hovering bird (seen from above in (a)), the wings sweep out an area known as the *wing disc*, whose area S_d is approximately equal to that of the circle whose diameter is the wing span. Air is sucked downward through the wing disc (b), and passes through it at the *induced velocity* V_i. The air continues to accelerate below the bird, and should eventually reach a downward velocity of $2V_i$.

In stationary hovering, all of the mass flow is due to the induced velocity itself, but in forward flight air passes through the wing disc as a result of the forward motion of the bird also. The faster the bird goes, the more air passes through the wing disc each second, and so the less induced velocity has to be imparted to it in order to generate the required force. The induced power, being proportional to the induced velocity, thus decreases progressively with increasing speed, in the approximately hyperbolic manner shown in Fig. 2–2a.

2.3   Parasite power

A bird's body contains the same basic organ systems as that of any other vertebrate, and is necessarily an object of substantial size. However well it is streamlined, it is bound to create some drag as it is propelled through the air, and a second component of power is needed to overcome this. This component is called *parasite power* because, unlike the induced power, it is not connected with generating lift, and remains obstinately present regardless of improvements in the wings.

The parasite (or body) *drag* is proportional to the cross-sectional area of the body, and to the square of the forward speed. The parasite *power*, being the parasite drag times the speed, therefore increases with the cube of the speed, producing a curve somewhat resembling a mirror image of the induced power curve (Fig. 2–2b). The flying bird has to produce both induced power and parasite power simultaneously. If we assume for the moment that no further power is needed, the sum of these two powers can be considered the power required by an 'ideal' bird. The result of adding together the induced and parasite powers is to produce a U-shaped curve (Fig. 2–2c) with a well defined minimum.

2.4 Profile power

The third main component of power, the profile power, is needed because the wings have to be flapped, and, like any solid object, create some drag as they go. As the relative air speed changes all the way along a flapping wing, the calculation of profile power is very complicated. To simplify matters, we shall regard it here as constant at all speeds, and equal to twice the minimum power of the ideal bird. It should be remembered that there is no theoretical basis for this simplification, but detailed calculations on the power requirements of the pigeon indicate that the approximation should be quite near the truth in practice. By using it, we can arrive at a reasonably accurate curve of power required versus speed, from which useful deductions about the performance of flying animals can be made.

2.5 The power curve

Figure 2–3 shows the curve obtained by adding together all three main components of power. The general U-shape follows from the opposite trends of induced and parasite power, and can be expected to apply to any sort of flying animal. Some very useful deductions can be made about the capabilities of a particular flying animal, if certain key points on its power curve can be located. There is now a sufficient theoretical and experimental basis for making such estimates, and formulae which can be used for the purpose are given in the Appendix.

The three most instructive points on the curve are:

(1) The power required P_{hov} at zero speed (hovering). Most birds, bats and insects can hover, at least briefly, and some are adapted for this type of flight.

(2) The minimum power P_{min}, and the speed V_{mp} at which it occurs. At this speed the bird can remain airborne for longest on a given amount of fuel, but if its object is to cover as much distance as possible for a given amount of fuel (as will generally be the case on migration), it must fly at a somewhat higher speed.

(3) The speed V_{mr} and power P_{mr} giving maximum range. Power is work done per unit time, and speed is distance travelled per unit time: to maximize range, we need to maximize distance travelled per unit work, that is, to find the smallest ratio of power to speed. The reader can

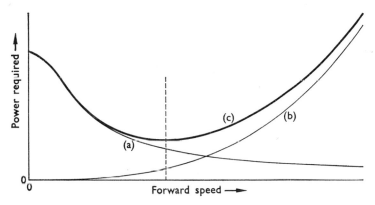

Fig. 2–2   The induced power (**a**) required to fly is a maximum at zero speed (hovering), and decreases as speed is increased, while the parasite power (**b**) starts at zero and increases with the cube of the speed. The sum of these two components (**c**) gives a U-shaped graph with a definite minimum.

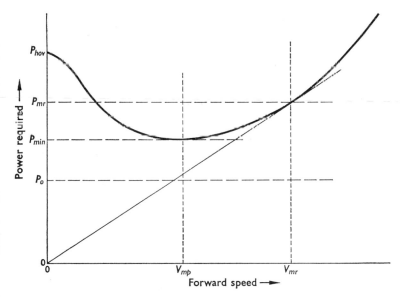

Fig. 2–3   Curve of power required versus forward speed, constructed by adding a constant profile power P_o to the curve of Fig. 2–2.

easily convince himself that this amounts to drawing a tangent from the origin to the curve of Fig. 2–3—the point where the tangent meets the curve gives the power and speed for best range.

The maximum range speed is at least 1.3 times the minimum power speed, and usually more—in the pigeon it is around 1.8 times the minimum power speed.

2.6 Relation of the power curve to body form

The induced power depends primarily on the disc area—the longer the wings in relation to the weight, the less induced power is required—while the parasite power is determined by the bulk and degree of streamlining of the body. Birds adapted to fly slowly or hover, like frigate birds, swifts, kites and terns, tend to have unusually long wings relative to their body weight, so reducing the induced power required, which is the largest component at low speeds. Birds like ducks, waders and pigeons, on the other hand, which are specialized more for fast cruising, have well streamlined bodies, so cutting down parasite power, the largest component at high speed. Since induced power is small at high speeds anyway, very long wings are of little advantage to a fast-flying bird—once it is off the ground. The short, broad wings seen in cursorial birds specializing in rapid takeoff, as in many Galliformes, allow the flight muscles to work at a high frequency, which in turn enables them to develop high power: this in itself, however, makes this type of adaptation suitable only for short bursts of activity, rather than for steady cruising (see Chapter 4).

2.7 Effective lift: drag ratio

In Chapter 1 the amount of power required to bring a gliding bird into horizontal flight was represented as a thrust force (equal to the drag), multiplied by the speed. A fixed-wing aeroplane corresponds quite closely to this simple arrangement, the thrust being an identifiable force supplied by a propeller or jet. In a flapping wing, however, the situation is much less clear, and one cannot really distinguish a particular component of force and call it 'thrust'. On the other hand one can define an 'effective thrust' from the relation that the power P equals effective thrust T' times speed V,

$$P = T'V$$

$$T' = \frac{P}{V} \tag{2.1}$$

If the speed and the mechanical power output from the muscles are known or can be estimated, then the 'effective thrust' T' can be defined

Plate 1 Two Australian flying marsupials of the family Phalangeridae. Like their placental counterparts, these animals glide from tree to tree, but they are not capable of powered flight. (**a**) The sugar squirrel *Petaurus norfolcensis* resembles the placental flying squirrels in the arrangement of its flying membrane (see also Fig. 1–4). (**b**) In the greater gliding possum *Schoinobates volans* the patagium extends only to the elbow. In flight the elbows are flexed, and the forefeet are held against the cheeks. The African scaly-tails (Rodentia; Anomaluridae) fly in a similar attitude, but they have in addition a cartilaginous strut which projects from the elbow joint and supports the leading edge of the patagium. (Photographs by Alan Root. Reproduced by courtesy of Tierbilder Okapia.)

(a)

(b)

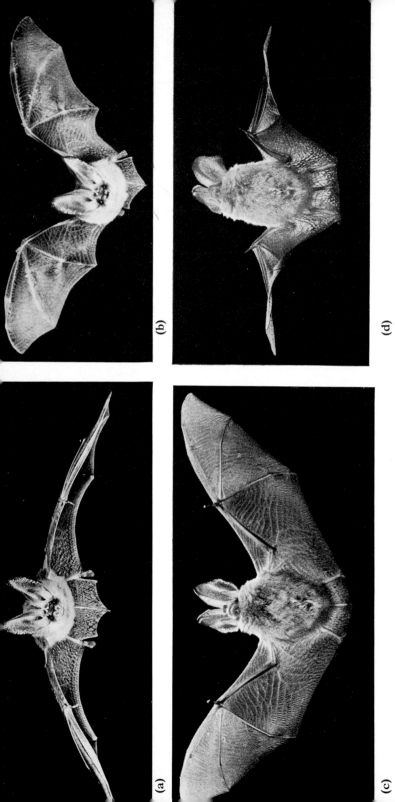

Plate 2 A long-eared bat *Plecotus auritus* (mass 9 g; wing span 26 cm) flying horizontally at 3·5 m s⁻¹. **(a)**, **(b)**: horizontal photographs from directly ahead. **(c)**, **(d)**: vertical photographs from directly below, simultaneous with **(a)** and **(b)** respectively.

(a), **(c)**: downstroke. The wing is fully extended and sharply cambered. The tail and feet maintain the curvature of the uropatagium and the proximal part of the plagiopatagium.

(b), **(d)**: upstroke. The proximal part of the wing has already been raised, and the distal part is being flicked to the 'fully up' position, inverting the wing tips momentarily.

The cycle of movements, analysed by Norberg (1971), corresponds closely to that described for the pigeon by Brown (1948), in spite of the differences as the ratio of power to speed, and in steady horizontal flight the effective thrust will equal the 'effective drag' D'. Similarly an 'average lift' L' must equal the weight W.

$$D' = T'$$
$$L' = W$$

We can now define an *effective lift:drag ratio* as the ratio L'/D'.

$$\left(\frac{L}{D}\right)' = \frac{W}{T'} = \frac{WV}{P} \tag{2.2}$$

This effective lift:drag ratio can be thought of as the ratio of the weight to the average horizontal force needed to propel the bird along. If no power is required, no horizontal propulsive force is needed, thus there is no drag and we have the perfect glider of section 1.3. In practice, however, power is always needed.

2.8 Range

A bird's range is the distance it can fly without refuelling. We have to distinguish between *still-air range* and *achieved range*, which is affected by the wind. In this section, still-air range will be considered.

If the work done in flying a small horizontal distance Y is E, then since work is force times distance, we can say that

$$E = YT' \tag{2.3}$$

T' being, as above, the average horizontal propulsive force exerted by the bird. From equation (2.1), this can be rewritten as

$$E = \frac{YP}{V} \tag{2.4}$$

If we regard E as some small amount of fuel used up, and Y as the corresponding distance flown, we can express equation (2.4) as

$$Y = \frac{EV}{P} \tag{2.5}$$

and from equation (2.2) this can be re-expressed in terms of the effective lift:drag ratio

$$Y = \frac{E}{W}\left(\frac{L}{D}\right)' \tag{2.6}$$

The quantity of energy E comes from the oxidation of a quantity of fuel which weighs something, and the ratio E/W in equation (2.6) may be regarded as specifying a certain fraction of the total body weight devoted to fuel. Equation (2.6) says that if a certain fraction of the body weight (1%, say) consists of fuel and is consumed, the distance the bird can fly depends *only* on the effective lift:drag ratio, not on the weight or the size of the bird. Thus if a hummingbird and a swan each start off with, say, 10% of their weight as fuel, we would expect both to go the same distance: if one in fact goes further than the other we should attribute this to greater aerodynamic efficiency (i.e. higher $(L/D)'$), not to any advantages resulting from large or small size as such.

In practice there is a general tendency for small birds to have somewhat lower effective lift:drag ratios than large ones, and also they cruise at lower speeds and hence are more affected by the wind. On the other hand, very large birds are restricted in the weight they can carry, so that the honours in long-distance non-stop migration tend to go to medium-sized birds, notably waders.

In spite of these provisos, it is generally true that small birds and bats are on a roughly equal footing with large ones in their ability to travel long distances non-stop. Combined with the ability to fly over water or any kind of terrain, this puts flying animals on an entirely different basis as migrators from walking or swimming ones, in both of which the distance that can be covered on a given fuel load (expressed as a fraction of body weight) is proportional to the linear dimensions, or to the one-third power of the weight. Thus one finds swans and hummingbirds making lengthy migrations, also caribou and whales, but not mice or minnows.

3.1 Morphological requirements for wings

In the evolution of any kind of flying animal three kinds of morphological changes have to take place if versatile and efficient powered flight is to be developed from simple forms of parachuting or gliding:

(1) The wings have to be lengthened, so increasing the area swept out when they flap. This is necessary to bring the induced power down to manageable levels for flight at low speeds.

(2) The muscular system has to be modified in such a way that it can transmit large amounts of power to the wings.

(3) The body has to be streamlined to make fast flight possible.

The lengthening of the wings is the most obvious anatomical adaptation to flight, and this feature alone leaves one in no doubt that the extinct pterosaurs were highly developed flying animals. In all three of the most highly evolved groups of flying vertebrates (pterosaurs, bats and birds), the need to develop long wings has led to drastic modifications of the original body form. The process has taken different routes in the three groups, and comparison between them gives an instructive insight into the principles involved.

Although the lengthening of the wings may seem a simple requirement in itself, it causes an awkward mechanical problem, in that the upward reaction on the wings, which must balance the animal's weight, has to act at some distance out from the body, resulting in a large bending moment at the wing root. Forces developed on the outer sections of the wings are of no use in flight if they merely bend the structure: they have to be transmitted to the shoulder joint, and thence to the body. Thus although the wing has to be thin for aerodynamical reasons, it also has to be very resistant to bending.

3.2 The pterosaur wing

The pterosaur solution to this problem (Fig. 3–1a) was to carry all the bending loads on a single bony rod. From the proximal end this consisted of the humerus, the radio-ulna, and then the metacarpals of digits 1–4, closely bound together. At the distal end of the metacarpals three clawed fingers projected forwards, and the wing 'spar' was then continued distally by the immensely elongated fourth digit, divided into four phalanges.

Some pterosaur fossils are so well preserved that the outline of the wing membrane can be seen. It was attached to the posterior side of the wing skeleton, and stretched across to join on to the side of the body, and the hind legs. Since the main wing surface apparently consisted only of a

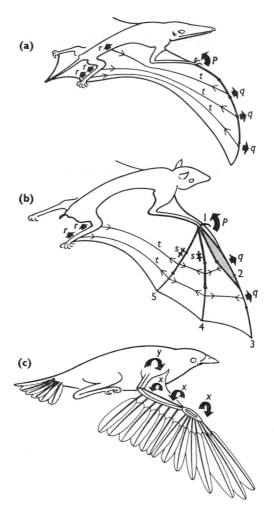

Fig. 3–1 (a) The pterosaur wing. The outward force stretching the patagium must have been supplied by a muscle whose action (*p*) protracted the fourth digit. The thin lines (*t*) represent tension paths in the membrane, between the forces (*q*) and (*r*) pulling outwards at the ends. In bats (b) the dactylopatagium minus (stippled), together with the parts of digits 2 and 3 surrounding it, constitutes a rigid unit, which is protracted by the pull (*p*) of the extensor carpi radialis longus muscle. The direction of the tension paths (*t*) in the membrane is changed by digits 4 and 5, which carry compression loads (*s*) because of this. This allows the forces (*q*) on the distal end of the membrane to act forwards rather than outwards, so that bats can have more pointed wings than would have been possible in pterosaurs. The numbers 1–5 identify the digits at their tips.

(*continued at foot of p. 17*)

fold of skin, having no bending strength in itself, it had to be suspended between at least *two* bony supports, and for this reason the hind legs also had to be modified as part of the wing structure, which must have severely limited the pterosaur's mobility on the ground.

Although the pterosaurs were a highly successful group for many millions of years, they died out along with the dinosaurs and many other groups at the end of the Mesozoic era, while the birds, which were few and inconspicuous at that time, survived and eventually replaced them. We can only speculate as to what caused the extinction of the pterosaurs in those days of general faunal impoverishment, but in comparison with the two modern groups of flying vertebrates, the birds and the bats, their wings showed an obvious defect in that the membrane had no skeletal support except at the anterior and proximal edges. In the first place this would mean that the membrane bulged upwards under air pressure, with no detailed control of its shape being possible, and secondly, a single tear would destroy the strength of the wing. The birds and the bats have each avoided these defects, but in different ways.

3.3   The bat wing

The bats, as an order of mammals, have, of course, no phylogenetic relationship with the pterosaurs, which were diapsid reptiles: the two lines must have been distinct since Permian, or even Carboniferous times, when the ancestors of both were somewhat lumbering quadrupeds. The wing structure in both groups is, however, based on the same mechanical principle. In the bats (Fig. 3–1b), as in the pterosaurs, the wing surface consists of a fold of skin which has only tensile strength and no resistance to bending, while the bending loads are carried by the skeleton of the limbs. The main difference is that in the bats there are four elongated fingers instead of one supporting the wing membrane. The leading edge of the hand wing is held forwards by a rigid but light unit, composed of digit 2, the proximal parts of digit 3, and the small piece of membrane (dactylopatagium minus) enclosed between them; the

---

The main surface of a bird's wing (c) is composed of the vanes of the flight feathers, and the aerodynamic force is transmitted through the bases of these feathers to the wing bones. The resulting tendency to twist the manus and ulna in the nose-down sense has to be resisted by supinating moments (x), applied via the proximal ends of the bones. Owing to the sharp angle at which the elbow joint is held in flight, the centre of lift lies ahead of the axis of the humerus: the resulting nose-up moment is counteracted by the pronating action (y) of the pectoralis muscle, which pulls downward on the deltoid crest, ahead of the axis of the bone. In contrast to bats and pterosaurs, a bird's legs are not involved in the wing structure, and can be modified independently for various forms of locomotion or for feeding.

mechanics of this arrangement have been analysed by NORBERG (1969). As in the pterosaur wing, the membrane is stretched between a finger (in this case digit 3) and the hind leg, but in the bat wing there are two more fingers, digits 4 and 5, which act as compression members, altering the direction of the tension paths in the membrane. They also serve to control the profile shape of the manus in a far more intricate way than is possible with the unjointed feathers of a bird, a feature which has given bats an unbelievable degree of manœuvrability in low-speed flight. The tension in the membrane is maintained by elastin fibres, and when these contract the skin crinkles, so that the wing area can be adjusted down to a minimum of about 75% of its maximum extent.

As in pterosaurs, the innermost panel of the bat wing is supported by the hind leg, which has the task of deflecting the central posterior part of the wing downwards in slow flight. In order to do this, the legs have had to be rotated at the thigh joints so that they are the other way round from those of most mammals—when a bat lies on the ground, its knees stick up posteriorly, in a grasshopper-like fashion. This greatly reduces the leg's effectiveness for walking, and while bats can clamber about with moderate agility, they are mostly poor at moving on a flat surface. When at rest they generally hang from their feet in the characteristic head-down posture, and take off by dropping from an elevated perch.

Both the bat and pterosaur arrangements could have been derived from a gliding arboreal animal with a simple patagium like a flying squirrel. The vital difference between them in the arrangement of the fingers could have arisen from a trivial distinction like that between the modern *Cynocephalus* which has all the fingers included in the membrane, and the flying squirrels where the membrane only extends to the wrists and ankles. These two forms are functionally almost identical at their present stage of development, but elongation of the wing in the latter would most likely lead to a pterosaur-like arrangement, whereas the former could give rise to the condition seen in bats.

3·4   The bird wing

The wing of a bird (Fig. 3–1c) is constructed in a radically different way from that of a pterosaur or a bat. The wing surface is made of feathers, which are thin, but stiff enough to be supported at one end only—instead of having to be stretched between two bones like the flexible membrane of a bat. The shafts of the feathers radiate out from the radio-ulna and the manus, and the forces and moments developed on the feathers are transmitted to these bones, and ultimately concentrated at the head of the humerus. One of the most important features of this arrangement is that the hind limbs are not connected with the wings, and are free to evolve quite separately for aquatic or terrestrial locomotion.

Like pterosaurs, birds were derived from bipedal diapsid ancestors, but unlike pterosaurs they have not had to give up their original bipedal method of locomotion in order to incorporate the hind legs into the wings. An outstanding feature of modern birds is that virtually all of them possess two entirely separate and independent locomotor systems, the wings in front which are usually adapted for flight (sometimes for swimming), and the legs behind, which may be adapted for either walking or swimming or both, besides being usually capable of grasping; the latter feature is used both in perching and in the manipulation of food. Land birds typically stand upright on their hind feet and walk or run about: by contrast, only a few bats, notably the vampires (Desmodontidae), are at all agile on the ground, and those that are walk on all fours.

3.5 Feathers

The flight feathers of a bird's wing—that is, the long feathers in the row which form the trailing edge of the wing—are unique structures adapted to collect the aerodynamic force distributed over the expanded vane, and transmit it as a concentrated bending moment at the base of the shaft (Fig. 3–2). One may imagine that the arboreal gliding ancestors of the birds would have developed enlarged scales, expanding the area of the arm posteriorly and distally, so reducing the gliding speed and flattening the gliding angle. Such scales would tend to curl up under the influence of the air pressure supporting the animal, and so developed a longitudinal stiffening ridge, now the rhachis. The vane would still tend to curl up at the edges, however, and corrugations in the vane were developed to resist this. These lateral ridges eventually separated to become the barbs, which are now separate structures linked together by a complicated mesh of little hooks, the barbules and barbicels. The vane constructed in this complex manner has a marvellous ability to withstand damage, and can reform an unbroken surface after being split by a blow. The intact vane is nearly, but not quite airtight, and it may well be that the controlled seepage of air through the feather is responsible for the remarkable high-lift properties of feathers, seen in hovering birds.

3.6 Mechanics of flapping flight

In level flight at a steady speed, the net result of flapping the wings must be to produce a force acting vertically upwards to balance the weight (Fig. 1–5). Considering a point half way along a wing in steady cruising flight, the relative airflow V_r which it 'sees' is the resultant of that due to the bird's forward speed (V) and that due to the flapping motion (V_f). Thus in the middle of the downstroke (Fig. 3–3a) the wing 'sees' a relative wind inclined upwards, so that the net aerodynamic force R acting on it (resultant of lift and drag) can be inclined forwards of the vertical, and provide a temporary propulsive thrust. At more distal points on the wing the vertical component of the relative airflow is larger, so that a larger component of forward thrust can be developed.

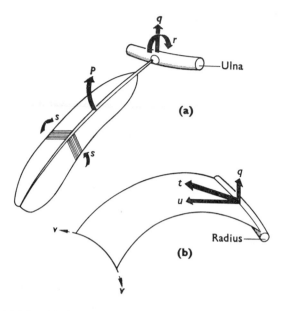

Fig. 3–2   (a) The lift force acting on the secondary feather of a bird tends to bend the feather upward (p). This tendency is resisted by the thickened rhachis of the feather, which passes the force through its base to the ulna as a combination of an upward force (q) and a moment (r) tending to pronate the bone. The lift also tends to curl the edges of the vanes upwards (s), and the ridges developed to resist this tendency have separated in modern birds to become the barbs.

(b) The force transmitted to the radius of a bat consists of the tension (t) in the membrane. Because the membrane bulges upwards this can be resolved into a vertical component (q), which provides the lift, and a horizontal one (u) which must be resisted by protraction of the radius. Because the lift does not apply a twisting moment to the radius, this bone in bats is relatively much thinner than the ulna in birds. The tension in the membrane is balanced at the posterior end by the pull (v) in a tendon along its curved edge.

On the upstroke, because the component of relative airspeed due to flapping is downward, the entire wing inevitably exerts a retarding force on the bird. The basic condition needed to make fast flapping flight feasible is thus that the angle of attack, and hence the magnitude of the net aerodynamic force, should be greater on the downstroke than on the upstroke. The effect of this can be clearly seen in still photographs of flying birds: on the downstroke the primary feathers are bent upwards and pressed firmly together, whilst on the upstroke they are deflected only a little or not at all, and may even separate.

At very low forward speeds, i.e., when the forward speed is small compared to the component due to flapping over most of the wing, most birds are able to obtain rather more useful effort out of the wing on the upstroke than they can in fast flight. Instead of raising the whole wing as

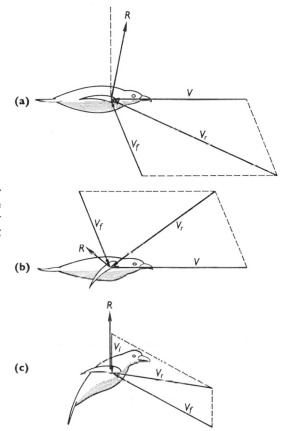

Fig. 3–3 Relative air flow and resultant force at a point midway along a flapping wing (see text).

a unit, the humerus is raised first, while the wrist joint is deflected in such a way that the manus remains depressed. The manus is then rapidly flicked into the 'fully up' position, in such a way that the primaries are inverted and generate an upward and forward force as they go. This complicated process was analysed by BROWN (1948) with the aid of high-speed photography, and his photographs are reproduced by GRAY (1968). A functionally similar wing action in a hovering bat has recently been described by NORBERG (1970b), (see Plate 2).

The hummingbirds (Trochilidae) are exceptional in that all the joints in their wing skeleton distal to the shoulder are fused, so that the elaborate movements seen on the upstroke in other birds are not possible. In a hovering hummingbird the wing, which is more or less symmetrical in cross-section, is simply inverted on the upstroke (or 'backstroke'): it is thus swept to and fro in a nearly horizontal plane, giving an upward force in both directions (Fig. 3–3c). In fast flight the action of the hummingbird wing is similar to that of other birds. An account of the hovering flight of hummingbirds, illustrated by series of stills from high-speed films, is given by GREENEWALT (1961).

3.7   Flight muscle system of flying vertebrates

The downstroke is the main work stroke in fast powered flight, and in all flying animals the depressor muscle of the wing is highly developed. In both birds and bats the pectoralis muscle performs this function, assisted to a small extent by certain smaller muscles. Its mass is generally about 15% of that of the whole animal, but varies from 10% to 22%, the higher values being found in animals which have high wing loadings for their size.

In both bats and birds the pectoralis has an extensive origin on the sternum, the rib cage, the clavicle and (in birds) the coracoid. It converges to insert on the ventral side of a shelf, which projects anteriorly from the proximal end of the humerus, and is called the deltoid crest in birds, and the pectoral ridge in bats. Thus besides depressing the humerus, the pectoralis also applies a moment tending to rotate it in the nose-down sense, which is balanced by the nose-up moment produced by the lift force on the wing.

Elevation of the forelimb in bats is performed by muscles on the dorsal side of the shoulder, in much the same way as in other mammals (see VAUGHAN, 1970). Birds, on the other hand, have a special elevator muscle for the wing, the supracoracoideus, which is arranged in a unique way, arising from the sternum parallel to, and completely covered by the pectoralis. Although its direction of pull is initially parallel to that of the pectoralis, its tendon passes over a pulley-like arrangement in the foramen triosseum (the canal where the scapula, clavicle and coracoid meet), and eventually inserts on the dorsal side of the head of the humerus.

The bird sternum has a prominent ventral carina or keel, which carries part of the origins of both the supracoracoideus and the pectoralis. Its function seems to be to support the pectoralis in such a way that it does not exert pressure on the supracoracoideus when it contracts, or occlude the branch of the interclavicular air sac which lies between the two muscles. The size of the carina is not necessarily related to the size of the pectoralis muscle in all flying vertebrates, and it is wrong to conclude that pterosaurs had feeble pectoralis muscles because most of them had only a poorly developed sternal keel. Most bats also have only a small keel, but the large pectoral muscles of the two sides have their origins in a median ligamentous sheet which extends ventrally from the

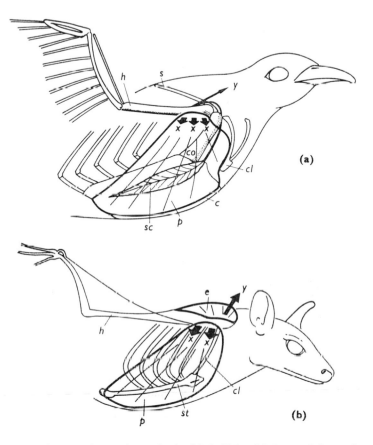

Fig. 3–4 The main work stroke in bird flight (**a**) is carried out by the pectoralis muscle (*p*) whose pull (*x*) depresses the humerus (*h*) and also tends to pronate it, because of the anterior insertion of the muscle on the deltoid crest of the humerus. The supracoracoideus muscle (*sc*), which elevates the humerus, lies underneath the pectoralis. Its tendon runs through the formamen triosseum, between the proximal ends of the scapula (*s*), coracoid (*co*) and clavicle (*cl*), and exerts an upward pull (*y*) on the dorsal side of the humerus.

In bats (**b**) the pectoralis is arranged much as in birds, but the elevator muscles (*e*) of the humerus have their origins dorsally on the scapula, which is much wider than in birds. The pectoralis muscles pull against one another, and are separated by a connective tissue sheet. The sternum (*st*) is quite small and does not have the large carina or keel (*c*) seen in birds.

sternum, and pull against one another (NORBERG, 1970a). This arrangement appears to be quite satisfactory in the absence of an underlying
antagonistic muscle, and of direct cooling of the muscles by air sacs.

The mass of the supracoracoideus muscle is a much more variable
fraction of the bird's mass than that of the pectoralis. Whereas the downstroke is a power stroke in both fast and slow flapping flight, the upstroke
only contributes a significant amount of power in slow flight (see section
3.6). In fast flight the upstroke is probably more or less passive, the wing
being raised by the lift force when the pull of the pectoralis is relaxed,
while in slow and hovering flight both propulsion and lift are obtained
on the upstroke. The supracoracoideus is therefore most strongly
developed in birds which are specialized for slow flapping flight, rapid
jump take-offs, or hovering. It reaches its greatest development in the
hummingbirds (Trochilidae), in which the ratio of the mass of the
pectoralis to that of the supracoracoideus is about 2:1, as compared to an
average of 10:1 for other birds (GREENEWALT, 1962).

3.8 Insect flight muscles

The main difference between the flight muscles of insects and those of
vertebrates is, of course, that they lie inside the skeleton. Apart from
this, some of the larger insects have muscles which are directly attached
to the proximal ends of the wings, as in vertebrates, but most have *indirect* flight muscles, whose primary effect is to distort the thorax. The
basic principle on which these work is shown in Fig. 3–5.

The mechanical effect of this arrangement is to 'gear up' the muscle
action, in such a way that a very small amount of shortening of the muscle
leads to a large angular excursion of the wings. Thus vertebrate skeletal
muscles typically shorten through 15–20% of their resting length on
their work stroke, whereas $1\frac{1}{2}$–3% is usual in the indirect flight muscles
of insects (PRINGLE, 1957). This very small proportional shortening is a
feature of 'fibrillar' muscles, which are found in the more advanced
flying insects, and can contract at much higher frequencies than ordinary
skeletal muscle (see section 5.7).

The wings themselves are flattened extensions of the cuticle, stiffened
by thickened ridges or 'veins'. Variation of wing area is generally not
possible, although some four-winged insects such as the Lepidoptera
can overlap the fore and hind-wings to a variable extent. The direction
and amplitude of the wing-beat are under muscular control, and so is
the wing profile shape in some insects.

3.9 Trim and stability in gliding birds

A bird or aeroplane is said to be 'trimmed' if it is flying straight at a
steady speed, with all the forces and moments acting on it in equilibrium.
'Stability' implies that if the equilibrium is disturbed, moments will be

set up tending to restore it. Both functions are achieved in most aeroplanes by the tail (Fig. 3–6).

Longitudinal or pitching trim is closely related to the control of speed. If, say, a downward force is applied and maintained at the rear end of an aeroplane by adjustment of the tailplane, the effect is to increase the angle of attack of the wing, and hence the lift coefficient, so

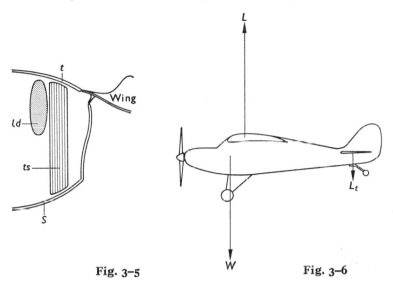

Fig. 3–5   W   Fig. 3–6

Fig. 3–5 (Left) Transverse section of one side of the thorax of an insect with indirect flight muscles. The tergo-sternal muscle (*ts*) compresses the thorax dorso-ventrally, so elevating the wing: at the same time distortion of the thorax causes lengthening of the antagonistic longitudinal dorsal muscle (seen in transverse section—*ld*), which depresses the wing when it contracts in turn (diagrammatic, modified after Snodgrass).

Fig. 3–6 (Right) In a conventional (tailed) aeroplane the lift on the wing (*L*) acts aft of the centre of gravity, producing a nose-down pitching moment, and this is balanced by a downward force (*L_t*) on the tailplane. Changes of trim are produced by adjusting the tailplane and hence L_t.

that a new equilibrium is reached at a lower speed (section 1.4). Once equilibrium is established, any rotation in pitch is opposed by the tailplane—a nose-up rotation, for example, produces an increase in the tailplane's angle of attack and hence an upward increment of force at the rear end, which tends to restore the original situation. It might be thought that the tail in birds would fulfil the same function, but in fact most birds (unlike aeroplanes) have no difficulty in flying without their tails. Trim in a gliding bird is effected instead by fore-and-aft movement of the wings, which can be rotated at the shoulder, elbow and wrist joints. To trim nose-down (faster) the wings are shifted back, and to trim nose-up (slower) they are moved forwards. It can be seen from Fig. 3–7 that the same movements which effect these trim changes also bring about a large reduction in the wing area at the higher speeds. The effect

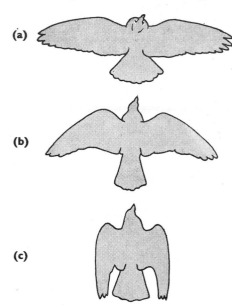

(a)

(b)

(c)

Fig. 3–7   A pigeon gliding in a wind tunnel at speeds of (a) 8.6 m s^{-1} (b) 12.4 m s^{-1} and (c) 22.1 m s^{-1}. The trim changes controlling speed are produced by fore-and-aft movement of the wings, which also result in a change of wing area over a range of 1.6:1, and of span over a range of 2.7:1. From PENNYCUICK (1968a).

of this is to make the change of lift coefficient with speed much less than equation (1.3) would suggest, and this in turn allows the bird to maintain a favourable gliding angle over a wider range of speeds than would otherwise be possible.

From the point of view of stability, a gliding bird or bat is comparable to a tailless aeroplane. Three ways in which such aeroplanes can be made stable are indicated in Fig. 3–8, together with suggested animal equivalents.

It is interesting that the earliest forms of both birds and pterosaurs (*Archaeopteryx* and *Rhamphorhynchus*, both from the Jurassic) had long tails which look well adapted to serve for control and stability as in aeroplanes. Later members of both groups dispensed with the tail, suggesting that the tailless method of control is more efficient aerodynamically, but more difficult to operate. It is not known whether the earliest bats also had long tails, since the earliest known fossil bat, *Icaronycteris* (Eocene), already looked much like a modern bat.

3.10   Use of the feet and the tail in flying vertebrates

The tail skeleton in birds is much reduced, and the visible tail consists of a fan of feathers (rectrices) which are similar in structure to the flight feathers. By fanwise spreading both the area and the span of the tail can be adjusted through a range which is typically about 3:1. As was mentioned in section 3.8, the tail is not the primary means of longitudinal control in birds, although it is undoubtedly used to supplement control movements of the wings, especially in rapid manœuvres. Its main function appears to be more analogous to that of flaps on aeroplane wings.

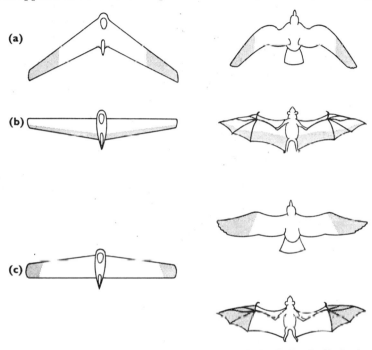

Fig. 3-8 Three methods of achieving longitudinal stability in tailless aeroplanes, with suggested animal equivalents. (a) Sweepback-with-washout: stippled areas twisted in the nose-down sense (e.g. Horten flying wings). Seen in birds when gliding fast. (b) Reflex camber: stippled areas deflected upwards (e.g. Fauvel flying wings). Seen in gliding bats. (c) Diffuser wing tips: stippled areas deflected downwards (e.g. Northrop flying wings). Seen in both bats and birds. From PENNYCUICK (1971c).

The tail is typically spread and depressed at very low speeds, especially at landing and take-off. This has two effects—in the first place some supplementary wing area is provided, and secondly the action of the tail helps to suck air downwards over the central portion of the wing, so increasing the maximum lift coefficient of the wing itself. Both effects enable the bird to remain airborne, and under control, at lower speeds than would otherwise be possible. In some birds with long forked tails, such as some swifts (Apodidae), swallows (Hirundinidae), terns (Sterninae) and frigate birds (Fregatidae), the effect of the tail when spread is to provide a long flap, preceded by a slot, posterior to the wing. Such birds are specialized for very slow flight and hovering, and use the spread tail for steep downward deflection of the airflow leaving the wings.

In bats the tail skeleton is relatively normal; it usually extends beyond the hind feet in Microchiroptera, but is very short in the Megachiroptera. In the former group the aerodynamical tail consists of a patagium stretching between the tail and the hind legs. This tail membrane is used at low speeds in basically the same way as in birds, but since the legs also support the inner end of the wing membrane the tail cannot be manipulated independently to the same extent as in birds. In most microchiropterans the area of the tail membrane is fixed, but in the free-tailed bats (Molossidae) the tail axis protrudes from the end of the membrane, and they can adjust their tail area by sliding the edge of the membrane towards or away from the tail tip (Fig. 3–9).

Fig. 3–9 In the Molossidae (free-tailed bats) the tail projects from the end of the uropatagium when the bat is at rest. In flight the edge of the membrane can be moved towards the tip of the tail (dotted line) so adjusting the tail area.

It has been remarked (section 3.4) that the independence of the legs from the wing mechanism in birds has enabled them, unlike other flying vertebrates, to develop the legs as separate walking, swimming and feeding organs. In many birds the legs have important functions in flight also, quite distinct from those of the wing. In gliding birds it is a basic requirement to make the lift:drag ratio as high as possible, but it is also necessary to be able to steepen the gliding angle to make controlled descents when required. Gliders are provided with airbrakes for this purpose, i.e., devices which produce drag but no lift, and this function in birds is performed mainly by the feet. The most effective airbrakes are the webbed feet of water birds, which can present a large flat surface perpendicular to the air flow. Vultures' feet are also highly effective as airbrakes, and interestingly enough the feet of Rüppell's griffon vultures *Gyps rüppellii* and white-backed vultures *Gyps africanus* have a small web between the middle and outer toes, which increases the effectiveness as airbrakes of the feet in these purely terrestrial birds (PENNYCUICK, 1971b), (see Plate 3).

In the auks (Alcidae) the tail is very short, and the large webbed feet can be used in a different way, as supplementary surfaces added to the sides of the tail in slow flight (Plate 3a). Just before touchdown the feet are rotated downwards from the lateral position at the sides of the tail, to absorb the deceleration of landing.

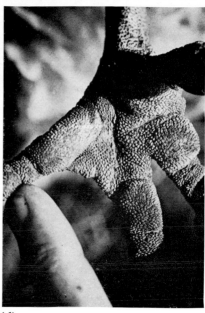

(d)

Plate 3 Use of the feet in flying birds. (**a**) The razorbill *Alca torda* has a high wing loading and a small tail. In slow flight it supplements its tail area by spreading its large webbed feet on either side of it. (**b**) A kittiwake *Rissa tridactyla* descending steeply, and using its feet for a different purpose: here the webs are spread at right angles to the air flow, and serve as highly effective airbrakes. (**c**) Land birds also use their feet as airbrakes in the same way. Here a lappet-faced vulture *Torgos tracheliotus* extends its legs and feet fully as it descends steeply on to a kill (**d**) The left foot of a white-backed vulture *Gyps africanus*, showing a small web between toes 3 and 4. This adaptation doubtless increases the effectiveness of the spread foot as an airbrake in this purely terrestrial species. (Photographs by C. J. Pennycuick)

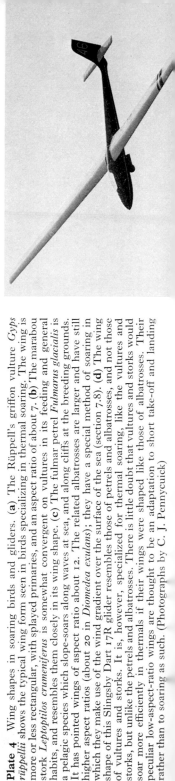

(c)

(b)

(a)

Plate 4 Wing shapes in soaring birds and gliders. (**a**) The Rüppell's griffon vulture *Gyps rüppellii* shows the typical wing form seen in birds specializing in thermal soaring. The wing is more or less rectangular, with splayed primaries, and an aspect ratio of about 7. (**b**) The marabou stork *Leptoptilos crumeniferus* is somewhat convergent on vultures in its feeding and general habits, and resembles them closely in its wing shape. (**c**) The fulmar petrel *Fulmarus glacialis* is a pelagic species which slope-soars along waves at sea, and along cliffs at the breeding grounds. It has pointed wings of aspect ratio about 12. The related albatrosses are larger and have still higher aspect ratios (about 20 in *Diomedea exulans*); they have a special method of soaring in which they make use of the wind gradient over the surface of the sea (section 7.8). (**d**) The wing shape of this Slingsby Dart 17R glider resembles those of petrels and albatrosses, and not those of vultures and storks. It is, however, specialized for thermal soaring, like the vultures and storks, but unlike the petrels and albatrosses. There is little doubt that vultures and storks would soar more efficiently in thermals if their wings were shaped like those of albatrosses. Their peculiar low-aspect-ratio wings are thought to be an adaptation to short take-off and landing rather than to soaring as such. (Photographs by C. J. Pennycuick)

4.1 Steady versus transient exertion

The work done by the flight muscles of a flying animal is derived from the oxidation of some fuel, either a carbohydrate or a fat. Of the chemical energy consumed, about one fifth appears as mechanical work, and the other four fifths as heat. In addition to the actual mechanical apparatus required for flight, the complete animal must have *supporting systems* capable of supplying fuel and oxygen to the flight muscles as required, and of disposing of excess heat.

One can distinguish broadly between the two types of locomotor activity—'cruise' and 'sprint'. In prolonged cruising flight oxygen must be absorbed in the respiratory organs and transferred to the flight muscles at the same rate as the muscles are using it up. The supporting systems—lungs and blood in a vertebrate—are thus presented by the muscles with a 'steady-state' demand, in which the rate of transfer of oxygen is the same as it passes from each stage to the next (lungs to blood, and blood to muscles). Similarly, fuel has to be removed from storage organs at the same rate as it is being used up, and heat has to be disposed of to the outside air at the rate at which the muscles produce it. These equalities determine the minimum capacities of the respiratory, circulatory and heat disposal systems needed to enable the animal to fly continuously.

Under 'sprint' conditions, however, the power output of the muscles temporarily exceeds the capacity of the supporting systems. Work is obtained quickly at the expense of depleting the small fuel reserves held in the muscle fibres themselves. Oxidation is anaerobic, resulting in an oxygen debt, which has to be repaid by subsequent increased oxygen consumption, and the extra heat is absorbed by a temporary rise in body temperature.

The ultimate maximum power output P_{max} which an animal can produce is determined by the strengths of the muscles and skeletal elements in its locomotor system, and not by the capacity of the supporting systems. The length of time for which power in excess of the capacity of these systems can be kept up, however, depends on the *reserve work* Q_r. This is the amount of mechanical work which can be obtained by mobilizing reserves, and it may be limited either by the amount of fuel stored in the muscles, or by the size of oxygen debt which can be accommodated, or by the maximum rise of body temperature which can be tolerated. If an animal exerts a power P which is greater than the maximum cruise power P_{ac} (i.e. the maximum power output with which the supporting systems can keep up), then this excess power can be maintained only until the reserve work has been used up. If this takes a time T_s, then

$$T_s = \frac{Q_r}{P - P_{ac}} \qquad (4.1)$$

Equation (4.1) applies to power outputs in the range P_{ac} to P_{max}. Powers above the latter limit are impracticable for mechanical reasons, while any power between zero and P_{ac} can be maintained indefinitely (not for a negative time as the equation suggests!). Equation (4.1) can be represented by the power-time curve shown in Fig. 4–1. This type of curve is best known from the work of BAINBRIDGE (1960) on fishes, where it appears in the form of a corresponding speed-time curve.

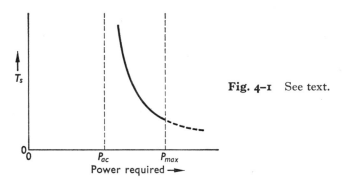

Fig. 4–1   See text.

Many terrestrial and aquatic animals pass their time in long periods of relative inactivity, punctuated by brief bursts of violent exertion. In this case the capacity of the supporting systems can be small in comparison with P_{max}. Flying animals, however, are typically capable of prolonged steady flight, calling for a relatively high sustained power output, and therefore need supporting systems of comparatively large capacity in relation to the animal's weight.

4.2   Blood system

The blood systems of flying vertebrates do not have any special anatomical features as compared with those of non-flying vertebrates, but there are some differences of proportion. For instance the weight of the heart in mammals other than bats is about 0.5% of the body weight, throughout the whole size range, while in bats and birds it is about 1% of the body weight. This presumably indicates that the total rate of blood flow needs to be relatively greater in flying than in non-flying animals, on account of the need to supply fuel and oxygen at high rates to the flight muscles.

4.3   Gas exchange and heat disposal

Birds have an entirely unique respiratory system in which the lungs themselves are small and non-distensible, and are connected to an extensive system of thin-walled, non-respiratory air sacs, which ramify throughout the viscera, and also penetrate the flight muscles and many of the bones. The system as a whole is tidal as in a mammal, but the lungs themselves have air channels (the parabronchi) which pass right through them—unlike mammal lungs, which are dead-end structures even in their respiratory portions. Many curious interpretations have been put on this arrangement, one being that the air sacs, being hollow, add lightness, and another that the bird lung is a 'through draught' system giving a very high rate of gas exchange.

The idea that the special arrangement of the avian lung is necessary for flight can be quickly dispelled by dissecting a bat, which proves to have lungs very similar to those of other mammals. Recent respirometric studies on birds flying in a wind tunnel by TUCKER (1968) have demonstrated further that bird lungs are able to extract up to 6.5% of the oxygen in the inspired air, and are not noticeably superior to mammal lungs in this respect.

Experiments designed to follow the movements of air flowing in the air sac system have given conflicting results, and it would seem that the route taken by air is most probably under the control of the bird to some extent. If this is the case, then the function of the air sacs as sites for evaporative cooling is readily comprehensible. Unlike mammals, birds do not have sweat glands, but it is clear from studies of thermoregulation in birds that heat can be disposed of at a high rate by evaporating water from the respiratory system (BARTHOLOMEW & CADE, 1963; SALT & ZEUTHEN, 1960). Unless some method of rapidly disposing of heat were available, it would not be feasible for birds to surround themselves with the efficient thermal insulation provided by their feathers, because of the danger of overheating during vigorous exertion.

Nearly all terrestrial birds and mammals rely on evaporative cooling, either in the respiratory system or by discharging fluid on the skin, as their primary channel for rapid disposal of large amounts of heat. An alternative method exists, however, namely convective cooling, whereby heat is transferred through the skin to a cool medium flowing past outside. Convective cooling is extremely effective in water and is the main method by which seals and whales lose heat, but in air it calls for some organ with a very large surface:volume ratio to be exposed to a rapid and continuous stream of air, at a temperature well below blood temperature. It so happens that the wing of a flying bat fulfils this specification, and bats are probably the only terrestrial vertebrates which can dispose of heat by convection fast enough to avoid overheating during prolonged and vigorous locomotion.

Heat disposal from the wings of *Myotis yumanensis* has been studied by REEDER and COWLES (1951). The wing membrane is richly supplied with blood vessels, but as its metabolic requirements are slight, only a small amount of blood is normally allowed to enter these vessels, in intermittent pulses. The amount can be increased if more heat has to be lost, and in a thermal emergency the whole of the wing plexus can be suddenly flooded with blood, but this only happens at extremely high body temperatures, 40–41 °C.

It is possible that the bats have rediscovered the earliest type of mammalian cooling mechanism, for many of the pelycosaurs of Carboniferous and Permian times had a huge 'sail' supported on the greatly elongated neural spines of their vertebrae. This was very probably a convective cooling organ, allowing these early mammal-like reptiles to develop thermal insulation before the introduction of the more convenient sweat gland.

The very thinness and lack of insulation which make the bat's wing so effective for heat disposal also make it rather a delicate organ, which is readily damaged by prolonged exposure to sunlight. This is probably the reason why bats have specialized on nocturnal or at least crepuscular flight, while the birds are predominantly diurnal. It is tempting to suppose that insectivorous bats fly at night because of the advantage conferred by their marvellous echo-location faculty, and avoid flying by day because of competition with insect-eating birds. This explanation would not apply to the tropical fruit-bats (Megachiroptera), which do not echolocate but also fly by night, although it is possible that danger from avian predators acts as an incentive to nocturnal flight in both groups.

Insects are generally thought of as poikilothermal animals, and so they are when at rest. In flight, however, so much heat is produced in the thorax of a large insect that some form of thermoregulation is necessary. HEINRICH (1970) has shown that in *Manduca sexta*, a large sphingid moth of mass 1.8–2.6 g, heat generated in the thorax is transferred by the blood to the abdomen, which is cooled by the airflow. The rate at which the blood is pumped is adjusted to keep the thoracic temperature at 41–42 °C, at any ambient temperature from 20–30 °C. If the flow of blood is stopped by tying the dorsal vessel, the moths cannot fly at ambient temperatures above 23 °C, unless the thick layer of scales covering the thorax is rubbed off, when they can tolerate air temperatures 7–8 °C higher. Thus the thorax is insulated, giving quick warm-up, and excess heat is dissipated as required through the thinly insulated wall of the abdomen. A functionally identical combination of endothermal temperature regulation in flight, with poikilothermy at rest, is found in most of the smaller bats (NOVICK and LEEN, 1969).

4.4 Fuels for flying animals

Both carbohydrates and fats are used as fuels by flying animals, and many flight muscles are specialized to run primarily on one or the other.

The energy yield from oxidizing a gram of fat is about twice that from a gram of anhydrous carbohydrate. However, carbohydrate has to be stored in a hydrated form, so that for a given amount of energy, the mass of fuel which has to be lifted in practice is about eight times as much for carbohydrate as for fat (Table 4.1). Thus for long-distance flights, where the maximum amount of energy must be obtained from a limited mass of fuel, fat is the more suitable fuel. It so happens that fat can only be utilized by aerobic oxidation in the muscle fibres themselves, so that a muscle running on fat cannot incur an oxygen debt, but this is immaterial in prolonged flight, where the muscles are operating under steady-state conditions in any case.

Table 4.1   Mass of isocaloric amounts of fuel stored by flying insects and birds. From WEIS-FOGH (1967).

Fuel	Water content (%)	Mass per unit energy (mg/cal)
Depot fat	0	0.11
Honey	20 (varies)	c. 0.33
Nectar	60 (varies)	c. 0.67
Glycogen	73	0.88

Muscles specialized for prolonged running on fat have to be richly supplied with mitochondria, and also usually contain myoglobin, which buffers the partial pressure of oxygen inside the fibres. Both of these components give the muscle a red colour. In vertebrates the fibres of such muscles tend to be small in diameter, with several blood capillaries running parallel to and in contact with each fibre, thus facilitating gas and fuel exchange between the blood and the interior of the fibre.

Muscles which are used only briefly for 'sprint' exertions generally run on glycogen, which is oxidized anaerobically at the expense of incurring an oxygen debt. Such muscles therefore do not need to contain mitochondria or myoglobin, and are white in colour, and generally larger in diameter than the red 'cruise' muscles.

The biochemistry of these two types of muscles, and their distribution in the flight muscles of birds and bats, is reviewed by GEORGE and BERGER (1966). Some birds such as the domestic pigeon Columba livia have fibres of both types intermingled in the pectoralis muscle. The red fibres are used for cruising and run under steady-state conditions, while the white fibres are used only during peak exertions, as at take-off. A few birds, notably game birds (Galliformes), normally fly only briefly when pursued and accordingly have a preponderance of white fibres in their flight muscles, but in most birds the red fibres predominate.

A sharp distinction between red 'cruise' and white 'sprint' muscles is not confined to birds, and an arrangement closely comparable to that seen in the pigeon is found in the myotome muscles of many fishes (BONE, 1966). Some confusion has survived from the earlier literature on this subject because in some terrestrial mammals, including man, the locomotor muscles are white while the slow muscles used in the control of posture are red, thus giving rise to the idea that white muscles are fast and red muscles slow. In fact the speed of contraction is a property of the contractile proteins and is not reflected in the colour of the muscle —the red colour being due to the presence of mitochondria and myoglobin. A red muscle, whether it is the slow postural muscle of a man or the fast flight muscle of a bird, is specialized for steady-state activity, whereas a white muscle is generally specialized for transient 'sprint' activity.

Like most birds, insects such as the larger Lepidoptera use fat as a fuel, and the locust *Schistocerca gregaria* uses carbohydrate for take-off and changes over later to fat (WEIS-FOGH, 1952) rather like a pigeon. Many of the smaller insects among the Diptera and Hymenoptera, however, run continuously on carbohydrate (PRINGLE, 1957). The muscles of flies like *Calliphora* and *Drosophila*, and bees such as *Apis* have very high specific power outputs because of their high contraction frequencies, and it may be that this in itself necessitates the use of carbohydrate. Alternatively, it may simply be that carbohydrate can be more quickly mobilized, and so is the more suitable fuel for the comparatively short flights in which these insects mostly engage.

4.5   Fatigue and endurance

It is often thought that prolonged exertion must in itself lead to exhaustion, but this is not necessarily so. Vertebrate hearts beat continuously throughout life, and there is no reason why locomotor muscles should not run continuously also, provided their demands for fuel, oxygen and heat disposal are within the capacity of the supporting systems. The duration of continuous flight may possibly be limited by such factors as the need for sleep, but the practical limit of endurance in migratory birds seems to be set simply by the amount of fuel which can be stored and lifted at take-off. There is good evidence at any rate that many small passerine species can fly for 50–60 hours continuously, and regularly do so on the trans-Saharan route (MOREAU, 1961).

Some ornithologists believe that swifts (Apodidae) fly continuously, day and night, and never land except when nesting. Whilst this has not been conclusively established, there is impressive circumstantial evidence that the European swift *Apus apus* actually does this (LOCKLEY, 1970). The swift's long wings in relation to its body mass would lead to an exceptionally low minimum power requirement in slow flight, and could well be an adaptation enabling it to keep flying all night on fuel saved up during the day.

5.1 Change of scale

If one changes the size of an object whilst keeping its shape the same, certain of its properties are changed in a predictable way. For instance if one takes a square drawn on a plain sheet of paper, and doubles the lengths of the sides, the area will increase by a factor of four: if the lengths of the sides are tripled, the area will be multiplied by nine, and so on. The same law applies to plane figures in any shape. If the figure is scaled up, so that all its linear dimensions are multiplied by n, then any chosen area will be multiplied by n^2 (Fig. 5-1).

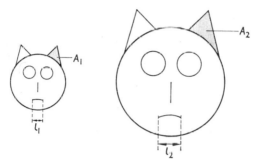

Fig. 5-1 If a plane figure of any shape is scaled up without changing the shape, then *any* chosen area increases in proportion to the square of *any* chosen length. Thus

$$\frac{A_2}{A_1} = \left(\frac{l_2}{l_1}\right)^2$$

This can be summarized by saying that for any plane figure, area (A) varies with length (l) squared. Expressed symbolically, this statement becomes

$$A \propto l^2$$

Similarly, for three-dimensional objects, *any* chosen volume (V) varies with the cube of *any* chosen length, or

$$V \propto l^3$$

Thus if potato A is three times the length of potato B, its volume will be 27 times that of B, and this is true regardless of the shape of the two potatoes, provided that both have exactly the same shape.

HILL (1950) applied this type of reasoning to animals and found that surprisingly far-reaching conclusions could be drawn about the effect of changes in size on locomotor performance. Hill was concerned mainly with walking and swimming animals, and some of his results are quoted here for comparison with those deduced by similar methods for flying animals.

One has to remember when applying simple scale arguments that the results are exactly true only for comparisons between *geometrically similar* animals, that is, animals which differ only in size, and are exact scale replicas of one another. In fact even closely related animals differ in the proportions as well as the sizes of their parts, and conclusions drawn from scale arguments are therefore at best approximate. Very large departures from geometrical similarity are, however, mechanically impracticable over a wide range of sizes, and scale arguments, although not extremely precise, give a good insight into the general nature of the effect of size on several interesting aspects of locomotor performance.

5.2   Effect of size on speed and on power required to fly

A bird's weight (W), being proportional to its volume, varies with the length cubed, while the wing area S, being an area, varies with length squared:

$$\left.\begin{array}{c} W \propto l^3 \\ S \propto l^2 \end{array}\right\} \quad (5.1)$$

It follows that the wing loading W/S varies directly with length:

$$\frac{W}{S} \propto \frac{l^3}{l^2} = l \qquad (5.2)$$

Flight at any characteristic speed, such as the speed for minimum power or maximum range, should occur at a constant lift coefficient throughout a series of geometrically similar animals. Thus if some such speed is referred to as V, it should vary with the square root of the wing loading (see section 1.4). Thus

$$V \propto \left(\frac{W}{S}\right)^{1/2}$$

so that, from proportionality (5.2)

$$V \propto l^{1/2} \qquad (5.3)$$

Proportionality (5.3) says that if bird A is geometrically similar to bird B but has n times the wing span (say), then it should go $\sqrt{n}$ times as fast. Notice that this result has been arrived at without having to calculate any particular speed, or even having to specify which particular characteristic speed is being considered. The identical result may be obtained by substitution in the formula for any characteristic speed, and the reader may like to verify this by starting from the formulae for V_{mp} or V_{mr} given in the Appendix.

For some purposes it may be more convenient to express such pro-
portionalities in terms of weight rather than length. In this case, since

$$l \propto W^{1/3}$$

Proportionality (5.3) can be written

$$V \propto (W^{1/3})^{1/2} = W^{1/6} \tag{5.3a}$$

Reverting now to the simple concept of horizontal flight represented
in Fig. 1–5, the muscle power P_r required to fly at any speed V is given by

$$P_r = TV \tag{5.4}$$

where T is a horizontal thrust force balancing the drag. It can be seen
from Fig. 1–5 that for any particular lift:drag ratio T must be proportional to the weight. If this, and proportionality (5.3) are substituted in
equation (5.4), we get

$$P_r \propto l^3 \times l^{1/2} = l^{3.5} \tag{5.5}$$

or alternatively

$$P_r \propto W^{1.17} \tag{5.5a}$$

Thus if bird A weighs twice as much as bird B it will require not twice
as much but $2^{1.17}$, or 2.25 times as much power to fly at its minimum
power speed (or alternatively at its maximum range speed). Since the
proportion of the weight devoted to flight muscle is much the same in all
flying vertebrates (GREENEWALT, 1962), this means that the larger birds
have to produce more power from each gram of muscle than do the smaller ones. This is so because larger birds are obliged to fly faster than
smaller ones, because of their higher wing loadings (proportionality 5.3).

The basal metabolic rates of different birds vary with about the two-
thirds power of the weight. Comparing this with proportionality (5.5a),
the ratio of power required in flight to the metabolic rate will clearly be
greater in large birds than in small ones. There is no basis for supposing
that this ratio would be constant in all birds, as some authors have as-
sumed. Of course, such comparisons only apply to birds which are flying
under corresponding conditions, and one cannot, for instance, compare
the oxygen consumption of a cruising pigeon directly with that of a
hovering hummingbird (as other authors have done).

5.3   Effect of size on power available

To see the significance of this finding we need to compare proportionality (5.5) with the corresponding relationship for the power available
from the flight muscles, P_a. P_a may be represented as the product of the
mass of the flight muscles m, and the power available from each gram of

muscle. This latter quantity is equal to the work done in one contraction by each gram of muscle (or specific work $\overline{Q}$) times the flapping frequency f. Thus

$$P_a = m\overline{Q}f \qquad (5.6)$$

m is, as has been remarked, approximately proportional to W, or to l^3. $\overline{Q}$ should be the same for any muscle able to exert a given stress and shorten through a given fraction of its rest length, and it may therefore be regarded as independent of l. f is bound to decrease with size for both mechanical and aerodynamical reasons (section 5.7)—that is bigger birds flap their wings more slowly than smaller ones. The maximum flapping frequency, limited by the strengths of the muscles and tendons, determines the maximum available power, and should vary inversely with length (HILL, 1950). Thus

$$\left.\begin{array}{c} m \propto l^3 \\ \overline{Q} \propto l^0 \\ f \propto l^{-1} \end{array}\right\} \qquad (5.7)$$

Referring to equation (5.6),

$$P_a \propto l^3 \times l^{-1} = l^2 \qquad (5.8)$$

or alternatively

$$P_a \propto W^{2/3} \qquad (5.8a)$$

Thus if bird A weighs twice as much as bird B it will need 2.25 times as much power to fly (section 5.2) but will only have $2^{2/3}$ or 1.59 times as much power available from its muscles. Evidently the *power margin*, or excess of power available over power required, must necessarily decrease with increasing size.

5.4   Maximum size of flying animals

Proportionality (5.5a) does not enable one to calculate the power required by any particular animal to fly. It says merely that if one plots a graph of the logarithm of the power required to fly versus the logarithm of the weight for a series of geometrically similar flying animals of different sizes, then the line should be straight, with a slope of 1.17. Similarly, proportionality (5.8a) says that the logarithm of the power available from the flight muscles should, when plotted against the logarithm of the weight, give a straight line with a slope of 0.667. In spite of the approximate nature of the argument, the difference between the two slopes is far too great to be evaded through departure from geometrical similarity, and the result is that there is a definite upper limit to the size and weight of practicable flying animals (Fig. 5–2).

Scale arguments, being inherently non-numerical, will not provide an estimate for the maximum practicable weight, but this maximum would appear on empirical grounds to be in the region of $120N$ for a bird with its crop empty, corresponding to a mass of about 12 kg. The largest members of several different orders weigh about this amount—for instance the Kori bustard *Ardeotis kori* (Gruiformes), the white pelican *Pelecanus onocrotalus* (Pelecaniformes), the mute swan *Cygnus olor* (Anseriformes) and the California condor *Gymnogyps californianus* (Falconiformes). It has to be remembered that the ability to fly horizontally is not the only requirement of an ecologically practicable flying bird. It must be able also to take off and climb fast enough to escape predators, to manœuvre safely close to the ground, and to carry a

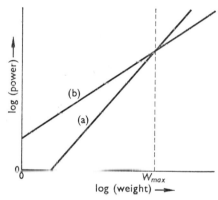

Fig. 5–2   (a) Power required to fly. (b) Power available from the muscles.

certain amount of food for itself, and usually also for its young. It is possible that special circumstances have eased one or more of these requirements at various times in the past, and thus allowed flying animals to exist which are larger than any alive today. BRAMWELL (1971) states that the largest known pterosaur *Pteranodon* (Cretaceous) had a probable mass of 18 kg, and estimates its performance on the assumption that it flew almost entirely by soaring. The same would apply to the huge super-condor *Teratornis* (estimated mass 20 kg) of the North American Pleistocene.

5.5   Take-off and landing

'Power required' has been discussed so far in this chapter as though it were a quantity characteristic of a particular animal, but in fact the power required depends of course on what the animal is doing. In horizontal flight power required is a function of speed, represented by the U-shaped curve discussed in Chapter 2. Power available, being independent of speed, can be plotted as a horizontal line on the same diagram (Fig. 5–3a). This diagram defines a range of speeds within which the power available exceeds the power required: the bird is therefore able to fly at any speed within this range, i.e. from V_a to V_b in Fig. 5–3a, but not at faster or slower speeds.

In practice two kinds of 'power available' have to be distinguished— the *maximum power available* P_{max} (in a brief sprint) which is determined by the mechanical properties of the muscles and skeleton, and the *maximum power continuously available* P_{ac}, which is determined by the capacity of the supporting systems (section 4.1). Figure 5–3b is a calculated

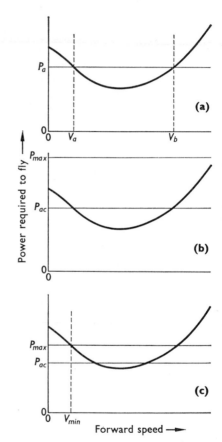

Fig. 5–3   See text.

required power curve for the pigeon, with estimates of P_{max} and P_{ac} marked in. It can be seen that the maximum power available is amply sufficient for flight at zero speed (hovering), so that the pigeon is able to take off from rest by jumping straight into the air. It cannot, however, hover continuously, since the continuous power available is sufficient only for flight in the range of approximately 3–16 metres per second.

If progressively larger birds are now considered, the power-required curve shifts upward and to the right, while the power available also increases, but not so much as the power required. Very large birds, such as the larger African vultures, have apparently only just sufficient power to fly horizontally, and cannot hover even very briefly. A bird with a power diagram such as that shown in Fig. 5–3c cannot fly at any speed slower than some minimum (marked V_{min}), and it must therefore accelerate to this speed by running along the ground before it can become airborne—unless, of course, it happens to be perched on a tree or hillside, in which case it may be able to attain V_{min} by an initial dive after take-off. The need for a take-off run is also decreased by taking off into wind, and if the wind strength equals or exceeds V_{min} no run is needed.

5.6 Load-carrying ability

If weight is added to a bird in the form of food carried in the talons, or stored in the crop or as fat, it is obliged to fly faster to achieve any particular condition of flight (such as minimum power or maximum range), and also to exert a larger effective thrust (section 2.7): for both reasons the power required to fly is increased. One consequence of the smaller power margin enjoyed by large birds as compared to small ones is that the amount of extra weight (relative to the unladen body weight) which can be carried decreases progressively as the size of the bird is increased. Many small birds can carry extra weight equalling their unladen weight, and migratory species commonly take on fat to about this amount before setting off on migration. There is little direct evidence as to which are the largest birds which can cope with fuel loads of this magnitude, but calculation suggests that birds of around 750 g unladen are about the largest which can double their weight and fly at their maximum range speed. Larger birds than this may still be able to lift a load equal to their unladen weight—as some large birds of prey no doubt can—but would then be restricted to speeds in the neighbourhood of the minimum power speed. In very large birds the amount of extra weight which can be carried is progressively restricted, until at a mass somewhere above 12 kg we arrive at the ultimate (but impracticable) bird which can just fly at its minimum power speed but cannot lift any extra weight at all. These limitations on weight-carrying ability have an important bearing on the theory of bird migration, and also on the evolutionary incentives for soaring, and these points will be taken up in Chapters 6 and 7 respectively.

5.7 Effect of size on wing-beat frequency

A bird flying forward at some particular speed can choose its wing-beat frequency over a range which has definite upper and lower limits. The upper limit f_{max} is determined by the force required at the insertions of the flight muscles to impart angular acceleration to the wing at the top and bottom of the stroke. When the muscles are exerting their maximum force at these points, no further increase of wing-beat frequency is possible. The lower limit f_{min} depends on the maximum lift coefficient of the wing (section 1.4). If scale arguments such as those used above are applied to these limits, it turns out that

$$\left. \begin{array}{l} f_{max} \propto l^{-1} \\ f_{min} \propto l^{-1/2} \end{array} \right\} \quad (5.9)$$

Thus as the size of a bird is increased, both maximum and minimum wing-beat frequencies decrease, but the maximum decreases more rapidly than the minimum. Ultimately there must be a size of bird at which the maximum and minimum wing-beat frequencies are equal at the minimum power speed, and this would set a different type of upper limit on the maximum practicable size of flying birds.

If one examines progressively smaller and smaller birds, their wing-beat rates get higher and higher, until in some of the smallest humming-birds (Trochilidae) wing-beat frequencies around 100 beats per second are found. Such very high wing-beat frequencies lead to another type of limitation, which is physiological rather than mechanical in origin. In vertebrate skeletal muscle contraction is initiated by an action potential, which propagates over the surface of the muscle fibre, and is itself initiated by the arrival of an impulse in the motor nerve. In one complete cycle of contraction the action potential has to travel over the muscle fibre, the excitation has to be transmitted to the contractile mechanism, the muscle has to shorten on its work stroke and lengthen again to its original length, and finally the whole system has to be reset in readiness for the next impulse. It seems that the time required for the complete cycle cannot be shortened much below 10 ms, so that contraction frequencies above 100 per second are not feasible in vertebrate skeletal muscle. A hummingbird with a mass of, say, 0.5 g would require a wing-beat frequency higher than this, and is therefore not a feasible type of bird. The body masses of both birds and bats show quite a sharp minimum in the region of 2 g.

Although the mass of the heaviest flying insect (*Goliathus goliathus*, Coleoptera) is about 40 g, and those of the lightest birds and bats are about 2 g, there are relatively few species of flying insects whose masses exceed 5 g, or of flying vertebrates below this. On the whole flying vertebrates range in mass from 12 kg down to 5 g, and below this mass the insects take over. With some exceptions, notably among the butterflies (Lepidoptera), insects mostly fly with wing-beat frequencies over 50 per second, and frequencies as high as 2000 per second have been recorded. Evidently insects have circumvented the wing-beat frequency barrier which limits the minimum size of flying vertebrates.

Most insects differ from flying vertebrates in that the nerve impulses to the flight muscles are not synchronized with the contractions of the muscles, and generally occur at a much lower frequency. The muscle contractions are instead triggered by mechanical changes within the muscle itself, and occur at a frequency which is determined by the mechanical resonance of the insect's wings and thorax. Muscles which are capable of this type of oscillatory contraction are known as 'fibrillar' muscles, and are only found as the flight muscles of insects, and as certain other specialized insect muscles, notably the tymbal muscles of the sound-producing organs of cicadas (PRINGLE, 1957). Not all insect muscles are of this type, however: some of the larger, slower-flapping insects, notably in the Lepidoptera and Odonata, have flight muscles of the one nerve impulse—one contraction type as in vertebrates. An insect with flight muscles of the latter 'neurogenic' type can be distinguished from one whose muscles are of the fibrillar or 'myogenic' type by clipping pieces off the wings, thus decreasing the moment of inertia and increasing the resonant frequency of the system. If the muscle rhythm is myogenic, the contraction frequency increases accordingly, whereas if it is neurogenic the contraction frequency remains the same, being determined by the central nervous system and not by the mechanical properties of the wings and thorax.

The mechanism of this fibrillar muscle leads to a very small proportional shortening on the work stroke, typically 1.5–3% of the resting length, an unusual mechanical feature which generally has to be accommodated by indirect insertions (section 3.8).

Migration

6.1 The physical problem of long-distance migration

The temperate regions of the earth, and especially the North Temperate Zone, provide a great variety of habitats for birds, but many of these are only usable for part of the year. Thus in England in the spring and summer there is (or was before the use of agricultural poisons became widespread) a huge abundance of insects, providing a rich but temporary food supply for such birds as swallows, wheatears, flycatchers, wagtails, various warblers and so forth. The biomass of insectivorous birds which the country can support is many times higher during the spring and summer months than during the autumn and winter. If the opportunities of spring and summer are to be fully exploited, then when autumn arrives most of the insect-eating birds must either change their diet, or become dormant (as do the insects on which they feed, and the bats with which they compete), or else move elsewhere.

For an insectivorous bird to use the northern European summer, and then move to an area of reliable food supply in time to avoid the winter, means in practice that it must travel south of the Sahara. The journey from England to Uganda, to take a typical example, spans about 50° of latitude, or roughly 5500 km, and the bird has 3–6 weeks in which to cover the distance, calling for an average rate of progress of 120–240 km per day. Such average speeds are out of the question for non-flying animals, quite apart from the problem of crossing the formidable barriers of the Mediterranean and the Sahara. Migration, and the use of seasonal habitats, as seen in birds today, is an ecological expedient which could never have developed without the prior evolution of the faculty of prolonged powered flight.

Most European bats hibernate rather than migrating in the winter, but some North American species are known to make regular migrations of a few hundred miles north and south. The distances covered are, however, not comparable to those traversed by birds on the regular long-distance routes.

6.2 Range attainable

It was pointed out in Chapter 2 that the distance a bird can fly by using up fuel comprising a given fraction of its unladen weight depends solely on its effective lift:drag ratio (equation 2.7). In long migratory flights the situation is complicated by the fact that the take-off weight may contain a very large proportion of disposable fuel, occasionally over 50%. As this fuel is used up, the bird's weight, and hence the power required to fly and the rate of using fuel, progressively decreases. In these circumstances, the actual distance Y flown is given by

$$Y = \frac{K}{g} \left(\frac{L}{D}\right)' \cdot \log_e \left(\frac{W_1}{W_2}\right) \tag{6.1}$$

where K is the amount of mechanical work obtained by oxidizing unit mass of fat (about 8×10^6 joules per kg), g is the acceleration due to gravity (9.81 m sec^{-2}), $(L/D)'$ is the effective lift:drag ratio, and W_1, W_2 are the bird's all-up weight at take-off and landing respectively. This formula assumes that the change in weight is due entirely to the consumption of fat.

Taking a typical $(L/D)'$ of, say, 6, and an extreme value of 2 for W_1/W_2, equation (6.1) yields 3400 km as an estimate of the maximum still-air range likely to be achievable in practice by migratory small birds. A more modest value of 1.5 for W_1/W_2 gives a range of 2000 km, which is probably more representative of the order of distance actually flown in a single stage by long-distance migrants. At a typical cruising speed for small passerines of, say, 35 km h^{-1}, a stage of 2000 km would take 55–60 hours. There is little doubt that many passerines do actually fly a stage of about this length when crossing the Sahara (MOREAU, 1961) although most of the less inhospitable migration routes can be subdivided into shorter stages.

These rough figures suggest that a small bird migrating from England to Uganda will need to spend approximately 160 hours (about one week) actually flying, but that this flying time will have to be broken into at least three stages with stops for refuelling in between. The total journey time will then exceed one week by whatever time is needed to replenish fat reserves at the intermediate stops.

A long stage such as the trans-Saharan crossing can only be accomplished reliably if the bird is able to feed well enough and quickly enough just beforehand to build up the very large fat reserve required. An important need in bird conservation is to protect areas which are used by migrants as refuelling stops, with the object of providing assured food supplies along known migration routes, at intervals which preferably should not exceed 1500 km. The destruction of refuelling areas by drainage of marshes or other human activities could easily lead to the extinction of migratory species, even when both their summer and winter quarters remain usable.

The longest overwater crossings by land birds are made by medium-sized waders of the Pacific, such as the bristle-thighed curlew *Numenius tahitiensis*, the Pacific golden plover *Pluvialis dominica fulva* and the ruddy turnstone *Arenaria interpres*, which cross from their breeding grounds in Alaska and the Aleutians to Hawaii and other northern Pacific islands. The shortest overwater crossing on this route—to Midway Is.—is about 2700 km, but there is some (admittedly inconclusive) evidence that longer direct crossings, up to 3700 km, are regularly made (JOHNSTON and MCFARLANE, 1967). Equation (6.1) indicates that a bird able to devote half its take-off mass to fat would need an effective lift: drag ratio of 6.5 (a modest value) to achieve the latter range, whilst if only one third of the take-off mass were fat, $(L/D)'$ would have to be 11.1, which is still not an unduly high figure for birds of this type.

JOHNSTON and MCFARLANE (1967) found that Pacific golden plovers shot on Wake Island in April contained rather small amounts of fat, ranging from 19–28% of their body mass. If these individuals were indeed about to set off direct for the Aleutians, the estimates of their effective lift: drag ratios would range from 21.8 (highly improbable) to 13.7 (possible, but rather high). One must conclude either that the birds normally stop at Midway Island, or that the pre-migratory build-up of fat is rapid, and that the birds shot had not yet reached their normal take-off mass. It is not unlikely that birds of this size could lift fuel up to 50% of their mass, and they would then have ample reserve fuel for the longest postulated crossings.

6.3   Effect of wind on range

The extent to which a bird's achieved range is affected by following or contrary winds depends upon the airspeed at which the bird normally cruises. For example a bird flying at an airspeed of 10 m s^{-1} against a head wind of 5 m s^{-1} will achieve a groundspeed of 5 m s^{-1}, so that it will cover only 1 metre over the ground for every 2 that it flies through the air. Its range will therefore be halved. A bird flying at an airspeed of 6 m s^{-1} under these circumstances will achieve only 1 m s^{-1} ground-speed, and its achieved range will therefore be one sixth of its still-air range, while a bird flying at an airspeed of 5 m s^{-1} would make no progress at all. Similarly a following wind produces a greater proportional increase in the range of a slow bird than in that of a fast one.

Figure 6–1 is a family of curves showing the effects of head and tail winds of different strengths on birds cruising at various speeds. The calculated cruising speeds (V_{mr}) of some birds, bats and insects are marked in, and the reader can calculate others from the formulae given in the Appendix.

The significance of this diagram is that although a small bird should be able to achieve the same *still-air* range as a large one, provided that its effective lift: drag ratio is the same (section 2.8), it flies more slowly and therefore needs to carry a bigger fuel reserve if it is to cover a stage of given length, with equal safety in the face of possible head winds. Flying speeds of insects are mostly too slow for long migration stages to be usefully reliable, and only a very few of the largest insects, such as the American milkweed butterfly *Danaus plexippus* make regular seasonal migrations of any length. Others, such as the locust *Schistocerca gregaria* rely on the wind rather than on their own motion through the air as a means of migration—a swarm of locusts drifts with the wind like an ani-mated balloon. The airspeed attainable by very tiny insects such as aphids (Aphididae) is an insignificant fraction of normal wind speeds, and such creatures rely entirely on the wind for their dispersal.

The effect of wind on migrating birds is actually not quite so great as the above argument would suggest, because they increase their air-speed when flying into wind, and decrease it when flying downwind. The reason for this procedure is easily understood if it is remembered

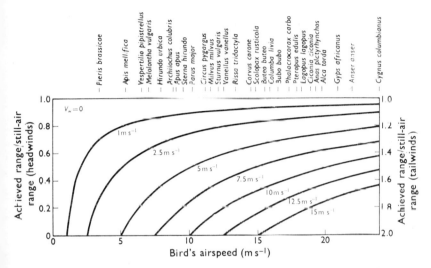

Fig. 6–1 Effect of head- and tailwinds on achieved range for eight windspeeds (V_w) from 0 to 15 m s^{-1}. The effect on range for an animal flying at a given airspeed is read off the scale at left for a headwind, or off that at right for a tailwind. Calculated cruising speeds (V_{mr}) for various animals are marked. From PENNYCUICK (1969).

that the cruising airspeed for maximum range can be found by drawing a tangent from the origin to the power curve, as in Fig. 2–3. If the bird's aim is to maximize distance travelled *over the ground* per unit work done, the graph must be redrawn with groundspeed instead of airspeed as the abscissa (Fig. 6–2). The effect of a tailwind of strength V_w is to shift the origin to the left by V_w, and the tangent now has to be drawn from the new origin, giving a lower *airspeed* than before for maximum range. Similarly, airspeed must be increased to obtain maximum range against a headwind.

Whilst the change of airspeed required is readily apparent once the wind speed is known, it is not at all clear how a migrant flying high at

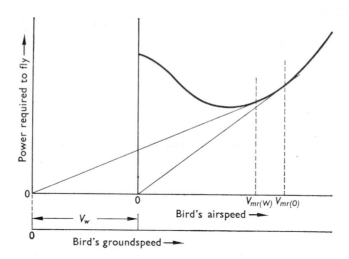

Fig. 6–2 When flying in a tailwind, a bird's groundspeed is equal to the sum of its airspeed and the windspeed. To construct a graph of power required to fly versus *groundspeed*, the y-axis of the power curve (Fig. 2-3) is therefore moved to the left by an amount equal to the windspeed (V_w). The airspeed giving the highest ratio of groundspeed to power (and hence maximum range) is found by drawing a tangent to the curve from the new origin. The airspeed so found ($V_{mr(w)}$) is slower than that for maximum range in zero wind ($V_{mr(0)}$). Similarly, when flying against a headwind, the bird must fly at an airspeed faster than ($V_{mr(0)}$) to achieve maximum range.

night could determine the windspeed accurately enough to enable it to take the necessary action. That birds do in fact adjust their cruising speeds in the appropriate way under these circumstances, however, has recently been shown in radar studies by BRUDERER (1971).

Soaring

7

7.1 Principles of soaring

In Chapter 2 the power requirements of a flying animal were discussed on the assumption that all the power would be provided by the animal's muscles. However, in certain circumstances some or all of these power requirements can be met by extracting energy from movements of the atmosphere. Flying in such a way as to extract energy from the atmosphere is called *soaring*.

Most forms of soaring involve manœuvring in such a way as to remain in an area of rising air. Suppose a bird with a certain lift:drag ratio (L/D) is gliding along at some speed V, it loses height at a vertical speed V_s (Fig. 7-1). If L/D is reasonably large, it is approximately true that

$$V_s = V\left(\frac{D}{L}\right) \tag{7.1}$$

V_s is termed the bird's *rate of sink*. If the air through which the bird is gliding happens to be rising at a speed greater than V_s, then the bird will be carried up with it.

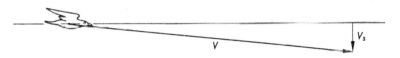

Fig. 7-1 See text.

Soaring is an opportunist method of flight, which depends on detecting and taking advantage of whatever vertical air movements happen to be available. Since the actual air is generally invisible, its motions have to be inferred from indirect evidence, and different soaring tactics have to be adopted to exploit 'lift' arising from different physical causes.

7.2 Slope soaring

The easiest method of soaring is slope soaring, in which the bird flies in a zone of rising air ('slope lift') caused by upward deflection of the wind by a hill. The area of lift is readily located as it remains stationary over the slope. The most usual manœuvre for exploiting slope lift is to tack back and forth above the steepest part of the slope, but if the wind strength exceeds the bird's minimum gliding speed it is often possible to hang motionless above a slope. Birds with low wing loadings (and hence low minimum gliding speeds) can often be seen doing this.

In winds of moderate strength (8–10 m s^{-1}) it is often possible to soar 100–300 m above the top of a soaring slope, but it is unusual to be able to maintain more than 500 m above a hill top using slope lift alone, regardless of the height of the hill itself. Slope lift is reliable in a steady wind blowing on to an isolated slope (such as a sea cliff in an on-shore wind), but if there are other hills to windward the slope lift may be greatly distorted or even non-existent.

Sea coasts often provide excellent soaring slopes facing prevailing winds, and many coast- and sea-birds are adept at slope soaring. A line of sand dunes will often provide sufficient lift for those of lower wing loading, such as gulls and terns (Laridae) to soar, while those species of gulls which frequent towns make full use of man-made slopes, such as esplanades and buildings, to patrol in search of scraps.

On British sea cliffs the fulmar petrel *Fulmarus glacialis* combines a passion for slope-soaring with a curiosity towards photographers, which has made it one of the most photographed of flying birds. In gale-force winds the auks (Alcidae), cormorants (*Phalacrocorax* spp.) and gannet *Sula bassana*—birds of higher wing loading than gulls or petrels—are able to soar also, and do so for hours on end. It is not very clear why cliff-nesting birds spend so much time slope soaring, and it could very well be that, like glider pilots, they simply enjoy soaring.

The sea birds with the higher wing loadings do not normally soar when away from the nesting cliffs, but the medium-sized and small petrels, which are entirely pelagic outside the breeding season, soar on the slopes of waves at sea. Far out to sea, it is sometimes even possible to do this in zero wind, by gliding along the forward slope of a wave which is moving relative to the stationary air.

Among land birds, many members of the crow family (Corvidae) will slope soar when conditions are suitable, again for no very obvious reason. Among birds of prey (Falconiformes), however, the use of slope-lift for patrolling a hillside in search of prey is very common, and the smaller members of this group, such as the kestrel *Falco tinnunculus* often use the slope lift along road and railway embankments in flat country for this purpose.

7.3   Thermal soaring

In spite of the widespread use of slope-soaring by birds, its limitations in respect of the limited height available, and in being confined to the windward slopes of hills, somewhat restrict its usefulness. In the warmer parts of the world, and especially in the interior of continents, *thermal soaring* provides much greater scope.

A *thermal* is a vortex structure caused by thermal instability in the atmosphere, and two broad types of thermal—the dust-devil type and the vortex ring type—can be distinguished.

The dust-devil, or columnar type of thermal, consists of a rapidly rotating column of air, which develops a low-pressure column in the middle owing to the rotation (Fig. 7–2a). Friction with the ground causes air to enter this low-pressure region at the bottom, resulting in an upward flow in the middle of the column. Such dust-devils are often rendered visible by dust and other miscellanea sucked up from the ground, hence the name. Dust-devils usually persist for only a few minutes and seldom provide lift over about 500–1000 m above the ground. Over an intense and persistent source of heat, however, such as an East African grass fire, a vigorous 'smoke-devil' may give good lift to as high as 2000 m above the ground for a longer period.

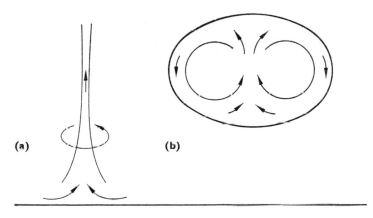

Fig. 7–2 (a) Dust-devil. (b) Vortex-ring.

Thermals of the vortex ring type (Fig. 7–2b) are sometimes triggered off directly from the ground by solar heating, or other sources of local heating, or they may arise in the atmosphere some distance above the ground. This type of thermal consists basically of a toroidal vortex with its plane horizontal, in which there is circulation of air up the middle and down the outside. The whole vortex gradually ascends and grows, and may have a life of half an hour or more. It is not uncommon to be able to climb 2500 m above the ground in thermals of this type, and at the higher levels the area of useful lift may grow to a kilometre or more in width.

Over the East African plains, on afternoons when soaring conditions are well established, it is common to find thermals of the dust-devil type predominating near the ground—up to 1000 m or so—giving strong, but narrow and short-lived patches of lift, whilst higher up a regularly spaced pattern of large vortex ring thermals may develop, often conveniently marked by blobs of cumulus cloud which form at the tops of the thermals. In both types of thermal the area of lift is roughly circular (rather than linear as in the case of slope lift), so that the appropriate soaring tactic is to fly round in steady circles, making corrective manœuvres as necessary to shift the circle into the strongest part of the lift.

7.4   Soaring as a means of food searching

Soaring as the basis of a specialized method of food searching is only possible in areas, and at times, where the occurrence of suitable weather conditions can be relied upon. It is most highly developed in birds of prey (Falconiformes) of tropical and sub-tropical areas, and especially in vultures—both the American vultures (Cathartidae) and the Old-World vultures (Aegypiinae), which evolved independently but are strongly convergent in structure and habits. The larger members of both groups approach the upper mass limit for birds (e.g. the California condor *Gymnogyps californianus*—12 kg; the lappet-faced vulture *Torgos tracheliotus*—7 kg; Rüppell's Griffon *Gyps rüppellii*—7 kg; the white-backed vulture *Gyps africanus*—5.5 kg), and therefore have a poor power margin (section 5.3). In addition, being scavengers, they rely for their food on the sporadic and unpredictable occurrence of dead animals over a wide area, and need to be able to remain airborne for several hours each day in order to attain a sufficiently high chance of finding enough food.

In the Serengeti National Park in Tanzania, which is typical of the savannah grasslands and woodlands in which these birds occur, the vultures take off as early as possible in the day and soar in slope lift if any is available, making powered sorties if any likely source of food is seen. As solar heating develops, they attempt to circle in any small patch of lift, and eventually are able to remain aloft in thermals, independently of hills and rocks. When food searching they do not generally climb above 500 m above the ground, but patrol to and fro at relatively low altitude over the most promising game concentrations. In the afternoon, however, vultures can be met with soaring right up to cloudbase, which is often as much as 2000 m above the ground. The crop in vultures bulges visibly when full, and birds seen soaring at great heights often have full crops, and are then not searching for food but travelling across country (section 7.5).

The aerodynamical requirements for soaring in large strong thermals are not particularly exacting, but the use of the narrow weak ones which occur at low altitudes early in the day imposes a need to be able to turn in very small circles, whilst maintaining a low rate of sink. In practice this requirement means that the wing-loading must be kept low, and this is no doubt a factor in the evolution of the typical broad wings of vultures— very similar to those of the marabou stork *Leptoptilos crumeniferus* which is strongly convergent on vultures in both structure and habits (see section 7.6).

7.5   Cross-country soaring

Owing to the migratory habits of the ungulates which provide their food, the African vultures are often obliged when rearing young to search for food 100 km or more from their nests. Having filled their crops they then have to return to the nesting site, and they do this by using thermals in a slightly different way, as a means of cross-country travel rather than simply as a means of remaining airborne. The principle of this is straightforward—the bird simply climbs in a thermal to some substantial height, such as 2000 m above the ground, then glides off straight in the desired direction, gradually losing height at its sinking speed V_s after it leaves the thermal. After gliding a few kilometres, and losing a few hundred metres of height, the vulture contacts another thermal, and circles up to repeat the process (Fig. 7–3). Vultures travelling across country typically climb 500–1000 m in each thermal, and glide 6–12 km in between. However, the simple situation of Fig. 7 3, i.e. distinct thermals separated by dead air, is often far from reality. Thermals are often not randomly distributed, but have a tendency to form into lines, and by the adroit use of such thermal 'streets' (at which vultures are adept) it is often possible to prolong the straight glides to 30 km or more, sometimes with little or no loss of height.

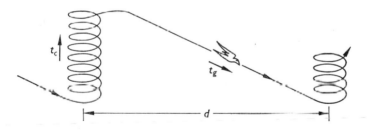

Fig. 7–3   See text.

The achieved groundspeed in simple cross-country soaring is given by the distance d from one thermal to the next, divided by the total time taken on the glide plus the climb ($t_g + t_c$—see Fig. 7–3), and is strongly dependent on the achieved rate of climb in thermals. Typically, large vultures can achieve cross-country speeds between 40 and 50 km h^{-1} in good thermal conditions, or somewhat over half their maximum range speed (V_{mr}—see section 2.5) in powered flight. In practice, of course, these large birds would not have sufficient muscle power to attain V_{mr} in level flight, and would have to go at some slower and less economical speed if obliged to travel under power.

Similar soaring techniques to those used by vultures for travelling between feeding and nesting areas are used for long-distance migration by white storks *Ciconia ciconia*, and certain migratory eagles (*Aquila* spp.), which nest in northern Europe and winter in East Africa or further south. These birds achieve their migration at a very low cost in energy, but have to accept certain limitations.

Although the bulk of the energy expenditure required for flight is avoided by soaring, basal metabolism remains, and the bird also has to expend energy in holding the wings down in the horizontal position. The latter would be expensive in energy if it had to be done by the fast muscles used for flapping flight, but it seems that in all birds which soar, whatever their systematic affinities, the pectoralis muscle is divided into a large superficial portion, and a small deep portion which acts synergistically with it, but has about one eighth to one twelfth the mass (KURODA, 1961). There is little doubt that this deep portion of the pectoralis, which is not found in non-soaring birds, is a specialized tonic muscle used to hold the wings in position during gliding.

Calculation of the energy consumption of these tonic muscles is somewhat conjectural, but rough calculations indicate that a large vulture or stork travelling across country in good thermal conditions should be able to achieve an average speed about equal to its minimum power speed (around 45 km/h), and that its total rate of energy expenditure would then be about one thirtieth as much as that needed to fly at the same speed under power. Thus a bird such as a white stork, migrating by soaring between northern Europe and tropical Africa, averaging, say, 6–7 hours' soaring per day, could expect to cover some 300 km per day, and would take 2½–3 weeks for the whole journey. This compares with the flight time of 1 week (day and night) calculated in section 6.2 for a passerine, but the difference is not really so great as these figures would suggest, because the soaring bird, owing to its much smaller energy expenditure, would need to spend little if any potential flying time refuelling along the route. The energy saved by thermal soaring on migration must, however, be offset against certain inconveniences. Progress depends on suitable weather, and it is not normally possible to fly at night. Also certain areas, notably the Mediterranean Sea, do not normally produce thermals, and have to be avoided by detours over land. It would appear that the advantages of soaring on migration outweigh the disadvantages for many large birds (storks, eagles), but not for the smaller ones, which perhaps find it easier to accumulate the necessary fat reserves for powered migration, and whose wing loadings would be too low to give them adequate soaring performance. Some large birds, notably geese and swans, also migrate by powered flight, though not on stages comparable in length with the trans-Saharan route.

White storks and white pelicans *Pelecanus onocrotalus* (which also migrate by thermal soaring, though not over such great distances) show an interesting form of group behaviour which increases the probability of finding the strongest lift, and hence the achieved cross-country speed. A flock of storks or pelicans, which may number several hundred individuals, spreads out horizontally when gliding between thermals, so that the flock as a whole scans a wide path in the search for the next thermal. As soon as one part of the flock flies into lift, and starts to rise visibly with respect to the rest, the others converge upon it, and if the thermal is strong the whole flock then quickly concentrates into a tight bunch circling together in the strongest part of the lift.

7.6 Gliding performance

'Performance' in horizontal powered flight is expressed by the curve of Fig. 2–3, which relates the rate of energy expenditure to the forward speed. In gliding flight the energy is provided by gravity, and the corresponding rate of energy expenditure is equal to the animal's weight times its rate of sink. A graph of rate of sink plotted against forward speed is called a *glide polar* (Fig. 7–4) and is the gliding equivalent of the power curve of Fig. 2–3. It is traditionally plotted downwards as in Fig. 7–4 because rate of sink is considered a negative vertical velocity.

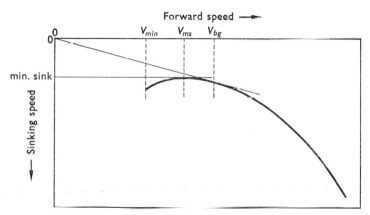

Fig. 7–4 Glide polar (see text).

Unlike a power curve, a glide polar does not extend to zero speed, but only down to the minimum gliding speed V_{min}, which is determined by the wing area and the maximum lift coefficient (section 1.4). Corresponding to the minimum power speed on the power curve is the *minimum sink* speed V_{ms}, while the (higher) speed V_{bg} for *best glide ratio* corresponds to the maximum range speed, and can be found in a similar way by drawing a tangent to the curve from the origin. The best L/D is found by dividing this speed by the corresponding rate of sink. Best glide ratios range from 10:1 to 15:1 in birds of prey and vultures, and may reach 24:1 in albatrosses. These figures are very poor in comparison with modern gliders, several of the best of which can achieve glide ratios exceeding 45:1.

The effect of increasing the weight or reducing the wing area is to increase both horizontal and vertical speeds in proportion to the square root of the wing loading. Thus birds with low wing loadings can glide slowly at low rates of sink, and are good at exploiting weak lift, while a high wing loading confers the ability to fly fast without excessive steepening of the gliding angle.

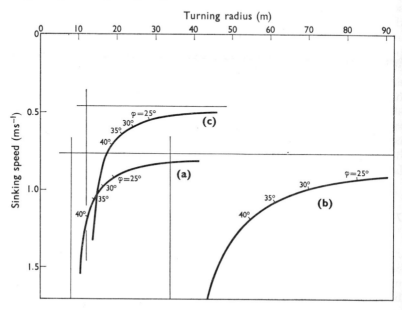

Fig. 7–5   Minimum sinking speed versus radius of turn for (a) the white-backed vulture, (b) the Schleicher ASK–14 powered sailplane, and (c) an 'albatross-shaped' vulture, which is an imaginary creature with the mass of a white-backed vulture and the proportions of a wandering albatross. As the angle of bank (ϕ) is increased the radius of turn becomes smaller, but the sinking speed increases. The horizontal asymptotes represent the minimum sinking speed in straight flight (infinite turning radius), which happens to be the same for the white-backed vulture and the ASK–14: the vertical asymptotes are the ultimate (but unattainable) circling radius, which would be obtained with 90° of bank. From PENNYCUICK (1971a).

Circling performance is the basis of climbing ability in thermals, and can be deduced from the straight-flight glide polar. It is usually represented by a graph of minimum rate of sink versus turning radius—the tighter the circle, the greater the minimum rate of sink. There is a minimum radius, tighter than which it is not possible to turn, and this radius is directly proportional to the wing loading.

Figure 7–5 shows the circling curve of the African white-backed vulture *Gyps africanus*, compared with that of a Schleicher ASK-14 powered sailplane which was used, with the engine stopped, as a flying observation platform to study it (PENNYCUICK, 1971a). The vulture, on account of its lower wing loading, can turn in much smaller circles than the glider, and is therefore better at staying airborne in small thermals, especially early in the day before larger ones have developed. On the other hand the glider's best glide ratio is nearly double that of the bird, so that it can achieve a much higher cross-country speed, so long as the thermals are big enough to sustain it. The third curve in Fig. 7–5 is for an imaginary bird with the same mass as a white-backed vulture, but the body proportions of a wandering albatross *Diomedea exulans*. This creature, if it existed, would outclimb the real vulture in all thermals except those with radii between 14 and 17 m, and would have a much higher maximum glide ratio in straight glides. It would seem that the short, broad wings typical of thermal soaring birds like vultures and storks are not an adaptation to soaring as such, but are more probably a compromise between acceptable gliding performance, and requirements dictated by some other factor, probably connected with take-off and landing capabilities.

7.7   Soaring in insects and small birds

After a few hours' thermal soaring on a summer's day in northern Europe, it is common to find the leading edges of a glider's wing encrusted with squashed aphids (Aphididae). There is evidence that these insects have a method of taking wing just as a newly-formed thermal breaks away from the ground, so that they are concentrated in, and carried up by the thermal. They then drift with the wind as the thermal drifts, and eventually fall to earth at some spot which may be many miles from the starting point. If the behaviour of these insects is specifically adapted to get them into thermals, then it should be classified as a primitive form of soaring. Locusts, by this criterion, do not really soar. Although SAYER (1962) has described how locust swarms are carried up to heights of 1800 m at the intertropical front over Somaliland, this seems to be quite accidental as far as the locusts are concerned.

European swifts *Apus apus* are also frequently encountered in thermals, and although they do not circle like bigger birds, but dart hither and thither, they nevertheless remain in the thermal, and are highly esteemed by glider pilots as indicators of the presence of lift. It is often thought that swifts are carried up in thermals more or less by accident, as a result of feeding on insects which become concentrated in the thermals by largely passive methods. Two observations suggest the contrary, however, that swifts are able to detect the presence of lift and actively remain in it. In the first place the habit is not shared by swallows (Hirundinidae) which are otherwise similar in their feeding habits; secondly, some species of African swifts, such as the little swift *Apus affinis* and the mottled swift *A. aequatorialis*, which nest in colonies on cliffs or isolated rocks, remain conspicuously concentrated in any patches of slope lift which may be present, when feeding or engaging in communal aerial displays. The swifts (Apodidae) thus have a claim to be considered the smallest soaring birds.

7.8 Dynamic soaring in a wind gradient

There is one method of soaring used by birds which does not depend on vertical air movements, but rather on variations in the horizontal wind speed. This is called dynamic soaring, and although the principle was understood long before the Wright brothers made the first soaring flights in 1902 (Lord RAYLEIGH, 1883), it still remains to be put into effect by glider pilots. In principle dynamic soaring is possible in random turbulence, but a more practical method makes use of *wind shear* over a flat surface, such as the sea. The layer of air in contact with the surface is slowed down by friction, and the full wind speed is only attained at some moderate height, like 50 m. The *wind gradient*, or rate of change of wind speed V_w with height h above the surface, is steep near the surface and tails off to zero further up (Fig. 7–6).

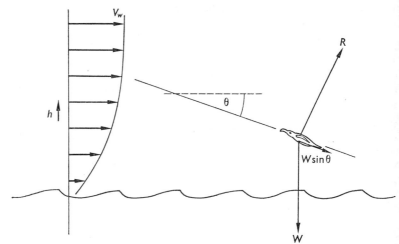

Fig. 7–6   The upwind climb in wind-gradient soaring (see text).

It has been realized for many years that albatrosses (Diomedeidae) are able to exploit this effect for soaring; their flight was studied by IDRAC (1924), and a very clear account is given by JAMESON (1958). First, as the bird heads into wind, climbing at an angle θ above the horizontal (Fig. 7–6), it tends to slow down because its drag, augmented by a component $W \sin \theta$ of its weight, acts backward along the flight path. On the other hand, a bird's usable kinetic energy depends on its airspeed, not its groundspeed, and as it climbs it gains airspeed, because the wind is blowing progressively faster towards it. As long as this gain is sufficient to offset the loss, the bird can continue to climb without loss of airspeed, and is able to soar. When the wind gradient becomes too weak to sustain any further climb, the bird turns downwind, and reversing the above argument then shows that it gains kinetic energy on the downwind glide also. Back at sea level, it soars in slope lift along the windward face of a wave until it meets a suitable upward gust, when it turns into wind and initiates another upwind climb. Albatrosses zigzag in this way over the Southern Ocean, progressing on average downwind, and circulate in the prevailing westerlies round and round Antarctica.

The rate at which the bird gains airspeed on the climb depends partly on the strength of the wind gradient and partly on the speed with which it climbs up through it. A high gliding speed (and hence a high wing loading), combined with a streamlined, low-drag body shape, is therefore best for this type of soaring.

It may be noted that albatrosses combine 'dynamic' soaring in the wind gradient with 'static' soaring along waves in rather an intricate way. The distinction between the two ('static' soaring depending on vertical air movements) is quite clear in principle, but not very useful in practice, because most practical soaring techniques involve elements of both kinds.

7.9 Other types of soaring

Two other types of soaring which are commonly used by glider pilots may be mentioned—front and wave soaring. Fronts form when two air masses of different properties meet with converging winds, and at some types of front, notably cold fronts and sea-breeze fronts, there may be a narrow but strong line of lift, sometimes hundreds of kilometres long. Some of the early studies of sea-breeze fronts in England, which result from incursion of sea air over the land in hot weather, were made by observing radar echoes from concentrations of swifts *Apus apus* soaring along the front.

Lee waves are standing waves which form on the *downwind* side of a hill in a steady wind. The upward sloping parts of the wave system may provide very strong lift, which, unlike slope lift, sometimes remains usable up to 10 times the height of the hill initiating it. Climbs to over 10 000 m above sea level have been made in several different countries by this method. Wave flying is one of the more difficult types of soaring, and little is known about the abilities of birds in this respect. Occasional sightings of soaring birds at great heights in mountainous areas indicate that they do sometimes make use of waves, but it is not known to what extent this is the result of regular behaviour or accident.

8.1 Types of methods

Although the flight of birds, in particular, has excited people's interest and admiration from the earliest times, the subject has proved resistant to scientific study because of the difficulty of making measurements on a free-flying animal. The solutions to this problem which have been devised fall into two broad categories—*free-flight* methods in which a bird at liberty is observed either from the ground or from an aircraft, and *controlled-flight* methods, in which known flight conditions are imposed on the bird, usually by means of a wind tunnel.

8.2 Free-flight measurements of gliding performance

Performance measurements on gliding birds are generally aimed at constructing a glide polar (section 7.6), which involves simultaneously measuring forward speed, and either rate of sink or gliding angle. A ground-based method of making such measurements devised by PENNY-CUICK (1960) took advantage of the behaviour of fulmar petrels *Fulmarus glacialis*, whose curiosity impels them to slope soar close past an observer in full view on a cliff top (the more conspicuous the better). A record of the bird's position during a pass was made with an optical tracking device, and later differentiated to give components of groundspeed along x, y and z axes. The readings of a three-dimensional anemometer held on a pole in the slope lift were then subtracted from these to give the corresponding airspeeds.

Most soaring birds are less obliging than the fulmar, and RASPET (1950) used an air-to-air method to measure the gliding performance of the American black vulture *Coragyps atratus*. The principle of this method is to fly along behind the bird in a glider, and measure the *difference* in speed, both horizontally and vertically, between the bird and the glider. The *glider's* forward speed is read off its airspeed indicator, and its rate of sink later found from a glide polar, constructed on the basis of previous calibration flights. Corresponding quantities for the bird are then obtained by addition.

Measurements of achieved cross-country speed in soaring birds can be obtained by simply following them about in a powered aircraft or glider (PENNYCUICK, 1972).

8.3 Wind tunnel measurements on gliding birds

A disadvantage of any free-flight method is that the bird is free to choose the speed at which it will fly, and so one ends up with a large number of measurements at the most commonly used speeds, and few or none at very high or low speeds. This can be overcome by training a bird to fly in a wind tunnel in such a way that it remains stationary relative to the experimenter (Fig. 8–1). If the tunnel can be tilted in such a way as to provide artificial slope lift, the flattest angle θ at which the bird can glide at any given speed V can readily be determined, and thus its glide polar can be constructed. This method also appears to have been first proposed by Raspet, although not put into effect by him. It has been used to obtain glide polars for a pigeon *Columba livia* (PENNYCUICK, 1968a), a falcon *Falco jugger* (TUCKER and PARROTT, 1970), a fruit-bat *Rousettus aegyptiacus* (PENNYCUICK, 1971c), and a Cathartid vulture *Coragyps atratus* (PARROTT, 1970).

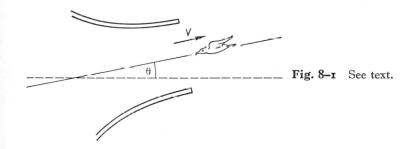

Fig. 8–1   See text.

8.4   Free-flight measurements on powered flight

In this category come simple measurements of the speed of free-flying animals, by 'pacing' them with a vehicle or timing them past two points with a stop-watch. Measurements of airspeed over long distances have been made by MICHENER and WALCOTT (1967), who attached radio transmitters to homing pigeons, and then tracked them with the aid of an aeroplane. Although these experiments were designed to investigate navigation, they have provided as a by-product some of the best measurements of sustained airspeed so far available.

HART and ROY (1966), and BERGER et al. (1970) used radio telemetry to measure the oxygen consumption of free-flying pigeons and other birds, in an attempt to tackle the more difficult question of measuring power consumption in free-flying birds. These measurements are difficult to interpret as yet, however, as results were obtained only for short periods of non-steady flight.

Another approach to this problem is to find the amount of fat which a bird consumes in flying a known distance. Unfortunately the fat content of a bird cannot be determined without killing it, but where a large population of small passerines is on the move, estimates can be made by sampling weights along a migration route. NISBET et al. (1963) used this method to estimate the amount of fat used by blackpoll warblers

Dendroica striata migrating south from the New England coast, and stopping at Bermuda, a distance of about 1200 km.

LEFEBVRE (1964) used an ingenious indirect method, based on measuring the relative turnover rates of isotopically labelled hydrogen and oxygen in water, to estimate the carbon dioxide output, and hence the metabolic rate, of pigeons *Columba livia* at rest and in flight.

8.5   Wind tunnel measurements on powered flight

Although not strictly a wind tunnel measurement, PEARSON'S (1950) measurement of the oxygen consumption of a hummingbird hovering under a bell jar deserves credit as the pioneer experiment in this field. This respirometric approach was extended by TUCKER (1966) who measured the oxygen consumption of a budgerigar *Melopsittacus undulatus* flying at various speeds in a small wind tunnel. This was done initially by sealing up the wind tunnel (which was of the closed-circuit type) and observing the rate at which the oxygen concentration inside declined. Later TUCKER (1968) improved the method and widened its scope by constructing a larger open-circuit wind tunnel, and measuring the bird's rate of oxygen consumption, and carbon dioxide and water output, by fitting it with a face mask and making the measurements externally.

A different type of method relies entirely on mechanical measurements, and the first experiment of this type appears to have been done by GREENEWALT (1961), who measured the maximum level-flight speeds of hummingbirds by training them to fly upwind to a feeder placed in the efflux of a variable-speed blower. This approach was developed by PENNYCUICK (1968b) to obtain a power curve for the pigeon *Columba livia*: by making measurements of the drag of the body, and of the amplitude and frequency of wing-beat at different speeds, the three components of power discussed in sections 2.1–2.4 were separately estimated, and then added together to yield the power curve. The information ultimately derived from this approach is basically the same as that from the respirometric approach, and there has so far been a gratifying degree of consistency between the results obtained by the two methods.

Appendix

It is often useful to make a rough, order-of-magnitude estimate of the speed at which a given animal should fly, or of the rate at which it will use fuel or oxygen. Such estimates can be made for a particular animal from the formulae given here, provided that at least its weight and wing span are known. A large collection of weights and measurements has been assembled from the literature by GREENEWALT (1962). The assumptions and approximations inherent in the formulae are set out by PENNY-CUICK (1969): it should be remembered that the calculations are approximate, especially when extrapolated to insects.

Units

SI units are used throughout in this book, and care must be taken to express all quantities in the right units before using the following formulae. The *weight* (W) must be expressed in the appropriate units of force, which are newtons. Weight cannot be expressed in kilograms, since these are units of mass, not force in the SI system. To convert a mass expressed in kilograms to the corresponding weight in newtons, multiply by 9.81, or by 10 for rough calculations.

Lengths must be expressed in metres, and *areas* in square metres.

Other kinds of units, such as those of the c.g.s. or f.p.s. systems, can, of course, be used without modification of the formulae, provided that units belonging to the same system are used consistently throughout.

Quantities required

The *disc area* (S_d) is the area of a circle whose diameter is equal to the wing span. The area actually swept out by the wings in flapping flight is irrelevant, and the full disc area should be used in all calculations.

The *equivalent flat plate area* (A) of the body is its greatest frontal area, multiplied by its drag coefficient. It is best obtained from wind tunnel tests on a frozen body, but if this is not practicable it can be estimated from the formula

$$A = (6.2 \times 10^{-4})W^{2/3} \qquad \text{(A.1)}$$

where W is the weight in newtons, and A is in square metres. This formula is appropriate for animals with well streamlined bodies, and A may be somewhat higher in species having long, trailing legs, large ears, and so on.

Air density (ρ) is a function of altitude, and can be obtained with sufficient accuracy by interpolation in Table A.1.

Table A.1   Air density as a function of altitude above mean sea level.

Altitude (m)	Air density ($kg\ m^{-3}$)
0	1.22
1000	1.11
2000	1.00
3000	0.905
4000	0.813
5000	0.732
6000	0.656

Characteristic speeds

The *minimum power speed* (V_{mp}) in metres per second is given by

$$V_{mp} = \frac{0.76 W^{1/2}}{\rho^{1/2} A^{1/4} S_d^{1/4}} \tag{A.2}$$

The *maximum range speed* (V_{mr}) is never less than 1.32 times the minimum power speed, and exceeds this by an amount depending on the profile power requirements (section 2.4). The evidence available to date suggests that

$$V_{mr} = 1.8 V_{mp} \tag{A.3}$$

should give a realistic estimate of the maximum range speed in most cases, and this is the airspeed at which birds may be expected to cruise on migration, subject to wind (section 6.3).

The assumption about profile power on which equation (A.3) is based is set out in the next paragraph: the formula can be adjusted to suit other assumptions by reference to Table A.2.

Power requirements

The first step in calculating power requirements is to find the absolute minimum power P_{am} (expressed in watts), where

$$P_{am} = 0.88 \frac{W^{3/2} A^{1/4}}{\rho^{1/2} S_d^{3/4}} \tag{A.4}$$

This would be the minimum power if no profile power were needed, but in practice profile power must be added. Although there is no simple means of doing this, a serviceable approximation based on studies of the pigeon *Columba livia* is that the profile power P_o is

$$P_o = 2P_{am} \tag{A.5}$$

The *minimum power required to fly* (P_{min}) (i.e. that required to fly at V_{mp}) is then

$$P_{min} = P_o + P_{am} \qquad (A.6)$$

The *power required to hover* (P_{hov}) is

$$P_{hov} = P_o + \sqrt{\frac{W^3}{2\rho S_d}} \qquad (A.7)$$

It should be realized that equation (A.5) is little better than a guess, and if better methods of calculating P_o become available in the future, then these should be used to find the value of P_o needed for substitution in equations (A.6) and (A.7).

The *power for maximum range* (P_{mr}) (i.e., the power required to fly at the most economical cruising speed V_{mr}) is about 1.3 times P_{min}, if equation (A.5) correctly represents the profile power. For other amounts of profile power, the ratio of P_{mr} to P_{min} can be read from Table A.2.

Fuel and oxygen consumption

The powers calculated from the above formulae represent the *mechanical* power required to fly, in watts. To estimate the rate of consumption of fat, in grams per hour (on the assumption that the muscles are 20% efficient) multiply this power by 0.45. To estimate the oxygen consumption in cubic centimetres per minute, multiply the mechanical power in watts by 15.

Table A.2 Effect of profile power ratio on performance estimates.

P_o/P_{am}	V_{mr}/V_{mp}	P_{mr}/P_{min}	$\dfrac{(L/D)'}{(L/D)_{ult}}$
0	1.32	1.14	1.00
0.5	1.45	1.19	0.706
1.0	1.57	1.23	0.556
1.5	1.68	1.26	0.465
2.0	1.78	1.28	0.403
2.5	1.88	1.30	0.357
3.0	1.96	1.32	0.323

Range

First estimate the effective lift:drag ratio $(L/D)'$ from the formula

$$\left(\frac{L}{D}\right)' = 0.4\sqrt{\frac{S_d}{A}} \qquad (A.8)$$

The formula assumes that the profile power is twice the absolute minimum power (equation **A.5**): for other ratios, see Table A.2.

Now let F be the *proportion* of the bird's take-off mass devoted to fuel (fat)—for instance, if half the take-off mass is fat, then $F=0.5$. The *air range* (Y) in metres is then

$$Y = (8 \times 10^5) \cdot \left(\frac{L}{D}\right)' \cdot \log_e \left(\frac{1}{1-F}\right) \qquad \text{(A.9)}$$

Divide this by 1000 to convert to kilometres. To assess the effect of wind, see section 6.3.

Note that an animal's weight need not be known to estimate its range (section 2.8).

References

BAINBRIDGE, R. (1960). Speed and stamina in three fish. *J. exp. Biol.*, **37**, 129–153.

BARTHOLOMEW, G. A. and CADE, T. J. (1963). The water economy of land birds. *Auk* **80**, 504–539.

BERGER, M., HART, J. S. and ROY, O. Z. (1970). Respiration, oxygen consumption and heart rate in some birds during rest and flight. *Z. vergl. Physiol.*, **66**, 201–214.

BONE, Q. (1966). On the function of the two types of myotomal muscle fibre in elasmobranch fish. *J. mar. biol. Ass. U.K.*, **46**, 321–349.

BRAMWELL, C. D. (1971). Aerodynamics of *Pteranodon. J. Linn. Soc. (Biol.)*, **3**, 313–328.

BROWN, R. H. J. (1948). The flight of birds: the flapping cycle of the pigeon. *J. exp. Biol.*, **25**, 322–333.

BRUDERER, B. (1971). Radarbeobachtungen über den Frühlingszug im Schweizerischen Mittelland. *Orn. Beob.*, **68**, 89–158.

GEORGE, J. C. and BERGER, A. J. (1966). *Avian myology.* Academic Press, New York.

GRAY, J. (1968). *Animal locomotion.* Weidenfeld & Nicolson, London.

GREENEWALT, C. H. (1961). *Hummingbirds.* Doubleday, New York.

GREENEWALT, C. H. (1962). Dimensional relationships for flying animals. *Smithson. misc. Collns.*, **144**, No. 2.

HART, J. S. and ROY, O. Z. (1966). Respiratory and cardiac responses to flight in pigeons. *Physiol. Zool.*, **39**, 291–306.

HEINRICH, B. (1970). Thoracic temperature stabilization by blood circulation in a free-flying moth. *Science, N.Y.*, **168**, 580–582.

HILL, A. V. (1950). The dimensions of animals and their muscular dynamics. *Science Progr.*, **38**, 209–230.

IDRAC, M. P. (1924). Étude théorique des manoeuvres des albatros par vent croissant avec l'altitude. *C. r. hebd. Séanc. Acad. Sci., Paris*, **179**, 1136–1139.

JAMESON, W. (1958). *The wandering albatross.* Hart-Davis, London.

JOHNSTON, D. W. and MCFARLANE, R. W. (1967). Migration and bio-energetics of flight in the Pacific golden plover. *Condor*, **69**, 156–168.

KURODA, N. (1961). A note on the pectoral muscles of birds. *Auk*, **78**, 261–263.

LEFEBVRE, E. A. (1964). The use of D_2O^{18} for measuring energy metabolism in *Columba livia* at rest and in flight. *Auk*, **81**, 403–416.

LOCKLEY, R. M. (1970). The most aerial bird in the world. *Animals*, **13**, No. 1, 4–7.

MICHENER, M. C. and WALCOTT, C. (1967). Homing of single pigeons—analysis of tracks. *J. exp. Biol.*, **47**, 99–131.

MOREAU, R. E. (1961). Problems of Mediterranean–Saharan migration. *Ibis*, **103a**, 373–427 and 580–623.

NISBET, I. C. T., DRURY, W. H. and BAIRD, J. (1963). Weight loss during migration. 1. Deposition and consumption of fat by the blackpoll warbler *Dendroica striata. Bird-Banding*, **34**, 107–138.

NORBERG, U. M. (1969). An arrangement giving a stiff leading edge to the hand wing in bats. *J. Mammal.*, **50**, 766–770.
NORBERG, U. M. (1970a). Functional osteology and myology of the wing of *Plecotus auritus* Linnaeus (Chiroptera). *Ark. Zool., Ser.* 2, **22**, No. 12.
NORBERG, U. M. (1970b). Hovering flight of *Plecotus auritus* Linnaeus. *Bijdr. Dierk.*, **40**, 62–66.
NOVICK, A. and LEEN, N. (1969). *The world of bats.* Edita, Lausanne.
PARROTT, G. C. (1970). Aerodynamics of gliding flight of a black vulture *Coragyps atratus. J. exp. Biol.*, **53**, 363–374.
PEARSON, O. P. (1950). The metabolism of hummingbirds. *Condor*, **52**, 145–152.
PENNYCUICK, C. J. (1960). Gliding flight of the fulmar petrel. *J. exp. Biol.*, **37**, 330–338.
PENNYCUICK, C. J. (1968a). A wind-tunnel study of gliding flight in the pigeon *Columba livia. J. exp. Biol.*, **49**, 509–526.
PENNYCUICK, C. J. (1968b). Power requirements for horizontal flight in the pigeon *Columba livia. J. exp. Biol.*, **49**, 527–555.
PENNYCUICK, C. J. (1969). The mechanics of bird migration. *Ibis*, **111**, 525–556.
PENNYCUICK, C. J. (1971a). Gliding flight of the white-backed vulture *Gyps africanus. J. exp. Biol.*, **55**, 13–38.
PENNYCUICK, C. J. (1971b). Control of gliding angle in Rüppell's griffon vulture *Gyps rüppellii. J. exp. Biol.*, **55**, 39–46.
PENNYCUICK, C. J. (1971c). Gliding flight of the dog-faced bat *Rousettus aegyptiacus* observed in a wind tunnel. *J. exp. Biol.*, **55**, 833–845.
PENNYCUICK, C. J. (1972). Soaring behaviour and performance of some East African birds, observed from a motor–glider. *Ibis*, **114**, 178–218.
PRINGLE, J. W. S. (1957). *Insect flight.* Cambridge University Press.
RASPET, A. (1950). Performance measurements of a soaring bird. *Gliding*, **1**, 145–151.
Lord RAYLEIGH (1883). The soaring of birds. *Nature, Lond.*, **27**, 534–535.
REEDER, W. G. and COWLES, R. B. (1951). Aspects of thermoregulation in bats. *J. Mammal.*, **32**, 389–403.
SALT, G. W. and ZEUTHEN, E. (1960). The respiratory system. Chapter X in *Biology and comparative physiology of birds.* Ed. by A. J. Marshall. Academic Press, New York.
SAYER, H. J. (1962). The desert locust and tropical convergence. *Nature, Lond.*, **194**, 330–336.
TUCKER, V. A. (1966). Oxygen consumption of a flying bird. *Science*, **154**, 150–151.
TUCKER, V. A. (1968). Respiratory exchange and evaporative water loss in the flying budgerigar. *J. exp. Biol.*, **48**, 67–87.
TUCKER, V. A. and PARROTT, G. C. (1970). Aerodynamics of gliding flight in a falcon and other birds. *J. exp. Biol.*, **52**, 345–367.
VAUGHAN, T. A. (1970). The muscular system. Chapter IV in *Biology of bats.* Ed. by W. A. Wimsatt. Academic Press, New York.
WEIS-FOGH, T. (1952). Fat combustion and metabolic rate of flying locusts (*Schistocerca gregaria* Forskal). *Phil. Trans. R. Soc. Ser. B*, **237**, 1–36.
WEIS-FOGH, T. (1967). Metabolism and weight economy in migrating animals, particularly birds and insects. In *Insects and Physiology.* Ed. by J. W. L. Beament and J. E. Treherne, 143–159. Oliver & Boyd, Edinburgh.